Thinking with the Dancing Brain

Thinking with the Dancing Brain

Embodying Neuroscience

Sandra C. Minton and Rima Faber

ROWMAN & LITTLEFIELD
Lanham • Boulder • New York • London

Published by Rowman & Littlefield
A wholly owned subsidiary of The Rowman & Littlefield Publishing Group, Inc.
4501 Forbes Boulevard, Suite 200, Lanham, Maryland 20706
www.rowman.com

Unit A, Whitacre Mews, 26-34 Stannary Street, London SE11 4AB

Copyright © 2016 by Sandra C. Minton and Rima Faber

All rights reserved. No part of this book may be reproduced in any form or by any electronic or mechanical means, including information storage and retrieval systems, without written permission from the publisher, except by a reviewer who may quote passages in a review.

British Library Cataloguing in Publication Information Available

Library of Congress Cataloging-in-Publication Data Available
ISBN: 978-1-4758-1250-3 (cloth : alk. paper)
ISBN: 978-1-4758-1251-0 (pbk. : alk. paper)
eISBN: 978-1-4758-1252-7

∞™ The paper used in this publication meets the minimum requirements of American National Standard for Information Sciences—Permanence of Paper for Printed Library Materials, ANSI/NISO Z39.48-1992.

Printed in the United States of America

This book is dedicated to:

My sister, Michele, whose stroke as a young woman catapulted us both into a study of the brain.

—RF

Dr. Alma M. Hawkins and Dr. Valerie Hunt, who ignited my interest in the subject matter, and Clarence Colburn, my husband, whose patience with the process was appreciated.

—SCM

Contents

Foreword by *Robert G. Shulman*	xi
Preface	xvii
Acknowledgments	xxi

1 Groundwork for Thinking in Dance — 1
- Introduction to the Brain — 1
- The Mind, Brain, and Body Connection — 4
- Somatic Practice — 5
- Laban Movement Analysis — 7
- Brain-Compatible Movement Patterns — 9
- Dance and Cognition — 10
- The Many Faces of Intelligence — 11
- Creativity — 13
- Learning and Movement — 16

2 Observation — 21
- The Human Optic System — 21
- Perception — 24
- Recognizing Patterns and Relationships — 27
- Nonverbal Communication — 30

3 Engagement — 37
- Brain Oxygenation — 37
- Attention — 39
- Motivation and Engagement — 44

4 High-Level Thinking Skills — 51
Categorization, Order, and Organization — 51
Critical-Thinking Skills — 54
Flexibility and Adaptability — 57
Self-Control and Self-Direction — 59

5 Emotions — 63
Emotions and Meaning — 63
Emotions and Learning — 68
Empathy and Learning — 71

6 Memory — 77
Memory Process — 77
Short-Term Memory — 80
Long-Term Memory — 82
Consciousness — 85
Spatial Memory — 88
The Emotion/Memory Link — 91
Recall — 92

7 Imagination and Imagery — 101
Imagination — 101
Imagery — 103
Visual Imagery — 105
Kinesthetic Imagery — 106
Body Image — 113

8 Learning — 119
Active Learning — 119
Embodied Cognition — 121
Patterns — 123
Associations — 125
Simile, Metaphor, and Analogy — 127
Transference of Learning — 130
Symbol Making — 132

9 Problem Solving — 139
Creative Problem Solving — 139
Logical Problem Solving — 142
The Solution — 144
Reflection — 146

10	**Twenty-First-Century Skills**	**151**
	Critical-Thinking Skills	152
	Communication	154
	Collaboration	156
	Creativity	158
Glossary		163
Index		171

Foreword

The role thought can play in dance flourishes to the extent that the behavior can be accurately and reproducibly analyzed. Faber and Minton plant their flag with their first words—"Movement is the embodiment of thought"—in which movement and thought work together in describing the person's response to the world. This knowledgeable book examines how behaviors, analyzed as movements in dance, are related to verbal descriptions that animate them and that they convey.

Faber and Minton's fusion of movement and thought is based on an understanding of dance as thoughtful movement. In their informed, impassioned fusion, the authors have resisted the reductionism offered by separating thought and movement.

In the visual cortex, they recognize that a myriad of neurons are involved in the learning and practice of dance. They trace how movement is handled by groups of sensory neurons that respond to the color, to the intensity, or to the movement of lines within the visual scene. This neuronal description of the visual input becomes a perception after modification by the observer's expectations. They describe active processes in which inputs from a visual signal or a dance movement are part of the localized brain processes of thought.

Thought does not cause the dance movement, nor does the movement cause the thought; any dualistic attempt to separate the behavior of dance from its thoughtful guidance by a thinking human is overwhelmed by the unity of thought and physical movement. The components of dance are assigned to the activities of certain neurons that are necessary for the person to have the thoughts, but they do not cause the thoughts, nor do they cause the actions.

The individual's mental expectations, expressed by delocalized activations throughout the brain, filter perceptions so that the significance of a thought is determined by memories, experiences, and the observer's history. The bonding of sensory input and mind is moving not from sensory inputs to thoughts but rather from memories and values, traditional constituents of the individual's history, to the perceptions of sensory inputs. As the authors say, "Nerves do not think by themselves. . . . The whole person imagines and creates."

A continuing direction in the book shows how sensory inputs, particularly the movements of dance, become the embodiment of thought, as first promised in the book's opening sentence. Faber and Minton's detailed appreciation of the movements of dance follows modern brain results by accepting some degree of localization of the brain activities accompanying the dance steps. They retain the primacy of movement by relating behavior to localized brain activities while including the importance of unexplored, delocalized brain activities.

Thinking and acting are inseparable, just as activities accompanying the movement have detailed aspects, describable by the teacher in a lesson, while at the same time the entire person contributes to the performance. In their parallelism of movements and thoughts, their detailed analysis of dance provides us with an enriched description of everyday behavior. Their text shows not how the movement is caused but how the thought processes are part of the behavior.

The extension of the concept of the mind to encompass the body and the world beyond has found favor in some recent philosophies. The question was raised by Clark and Chalmers,[1] who asked, "Where does the mind stop and the rest of the world begin?" The extent of mind was focused by Bateson,[2] who asked, "Where does the blind man's self begin? At the tip of the stick? At the handle of the stick? Or at some point halfway up the stick?"

The degree to which the body, with its musculature and hormonal responses that arise from the brain and which in turn activate the brain, is extended by considering the world beyond as not just the brain, not just the body, but all of this and beyond the body. In archeological studies of early humans, where the remnants consist of hand-hewn tools, Lambros Malafouris[3] has asked how the material world is responsible for the functioning usually attributed to the mind. These considerations have generated questions that were prohibited by the dualism in which thought is mapped onto the brain, questions that the present book resolves about the role in movement of thought.

We are well on to fusing the physical world with the brain when we have a degree of understanding such that as described by Faber and Minton in their analysis of teaching dance as a mental and mechanical process. Their conviction that mental processes are inextricable from movement when teaching dance has led them to relate teachable components in terms of mental activities that are traditionally assigned to the brain.

The authors find a role for built-in representations of the world, such as working memory, not as the cause of movements, but as part of the inextricable flow of bodily movements and intentions that form dance. They discuss dance both from its proposed origin in mind with cognitive capabilities and from the viewpoint of a bodily movement that flows inseparable from human life. But in their hands, cognitive concepts serve not to explain movement, which has led to oversimplifications and overinterpretations of results, but to remind us that thought is inseparable from movement, illustrated here by the many ways of teaching dance.

The present book, in proposing to teach dance, relates their analysis of behavior in the form of the movements required to dance to insights of neuroscience. Dance is the phenomenon, the effect that needs no cause. Our appreciation of the effect can be increased by tracing its background, as if it were created by a mind, but our understanding is created by the observed movements of dance, a detailed behavior that has a harder reality than any cognitive causes proposed. Their book shows an experienced appreciation of the phenomena, of the processes that are the essence of dance and of movement itself.

The authors refer to generally accepted causes of events, but their interests, in dance and in teaching it, bring them to full appreciation of dance itself. This appreciation has brought the elements of dance, as an intensely studied and understood component of behavior, into recognizable patterns. In my opinion, the author's elements of dance have a better chance of being reproducible and therefore capable of scientific explanation than do the concepts of cognitive psychology like memory or attention, presently valued in brain-imaging experiments.

Cognitive scientists have proposed that localized brain activity can provide scientific understanding of the concepts of thought. In relating the behavioral elements of dance to popular cognitive concepts, the present authors have stooped to conquer. In my opinion, their clear decomposition of dance movements is more relevant to the understanding sought by scientific studies than are the barren generalizations of cognitive concepts that are being claimed, in brain-imaging studies, as the route to such understanding.

By their breakdown of dance into teachable features, into actions involving muscle, bone, and training, they have created a detailed study of behavior, aided by limiting their studies to dance. In seeking the relations between cause and effect, so as to explain behavior, modern cognitive science has assumed that behavior—the effect—is readily understood, and that finding causes are the challenge.

In scholarly circles, behavior is usually described by psychological interpretation as studies of the causative powers of mental processes. This division of labor divides behavior from mental processes and then looks for connections, assuming that mental causes differ from behavioral effects.

The analysis, originating in Descartes's need to separate the physical from the spiritual, the religious from the secular, cannot integrate the dualism that drove the separation.

The book's multileveled analysis of behavior in dance reflects a choice between two prevailing ways of regarding brain activity. The first, a view that I am treating favorably and which is beautifully illustrated by Faber and Minton, is an empirical study of behavior linking it to the variety of human properties. In this view, the brain is plastic, responding to individual experiences, molded by history, and dependent on person, place, and specific conditions. There are no generalizations with universal applicability, but only understanding empirically derived in the particular case at hand.

Our unique understanding depends on our knowledge of behavior from which we can infer the tentative existence of concepts. The value of these concepts does not exist intrinsically in stating a generalization. Rather, their value depends upon their usefulness that is judged pragmatically by the effect its existence would have upon human activities. So our ideas about the dance movement do not have universal significance, but they do influence how we think of the behavior, and how we explain it to others as, for example, when teaching. The value of this book lies in its description of how thought, movement, and psychology interact when explaining it to others.

The other view, which contrasted with the empirical studies of the present book, treats the conceptualization of mental processes as identifiable cognitive concepts, like the ship at sea, different from its supporting medium. It postulates that there are mental processes that reside as distinct, identifiable modules in the brain that stand ready to be applied to a phenomenon. They are identified and their properties determined from cognitive psychology.

These concepts, based on a computerlike brain, are innate brain activities that are recognized as the underlying causes of observable behavior. In following this philosophy of brain functional imaging, experiments have been, in great part, dedicated to observing the same brain activation when different individuals, facing different examples of a memory task, are presumed to find activities in the same brain region. One of the great conclusions of years of searching for the same brain region being activated during a behavior characterized by these concepts is that no such brain regions have been identified. There is no agreement as to where working memory, or attention, or inference is located in human brains, nor can brain regions be located as the cause of related behavior, despite vigorous claims.

Dance is not the effect of some brain activity; it is the activity of interest because it is the stuff of life. It is a specific form of behavior whose well-developed, normalized structure allows a more controlled understanding of behavior than do the psychological concepts that dominate scientific thinking based on cognition.

The authors are too modest in relating their mechanisms of dance movement to the generalized cognitive concepts of contemporary interest. They discuss the relations of behavior with cognitive concepts into which cognitive science classifies perceptions—in deference to the pride of place that science holds. But dance has not been caused by neuronal activities or psychological concepts waiting to be triggered by an event. Rather, it is the lively nature of this broadly human event and the authors' abilities to teach it using communicable concepts that attest to its nature.

The authors refuse to accept the secondary position, relative to a causative mind, attributed to the movements and activities required for dance (and for living processes). They do not think that the explanation of what causes these movements is a higher understanding. Rather, they have described that teaching expands how we learn and how we profit from the interpretations offered as causes of movement and dance.

In contrast to settling for neuronal explanations of behavior, or for mechanical descriptions of movement, this book celebrates the thoughtful behavior of dance without muddying our experience with simplistic claims for causation. It is pleasing that the analysis of behavior in dance can liberate science from the barren concepts of cognitive psychology.

Robert G. Shulman, PhD
Sterling Professor (Emeritus) of Molecular Biophysics & Biochemistry
Yale University

NOTES

1. Andy Clark and David Chalmers, "The Extended Mind," *Analysis* 58:7–19 (Oxford: Oxford University Press, 1998).

2. Gregory Bateson, *Steps to an Ecology of Mind: Collected Essays in Anthropology, Psychiatry, Evolution, and Epistemology* (Chicago: University of Chicago Press, 1972).

3. Lambros Malafouris, *How Things Shape the Mind* (Cambridge: MA: MIT Press, 2013).

Preface

When Yvonne Rainer choreographed "The Mind Is a Muscle,"[1] she was referring to the mental gymnastics required to learn complex dance movements. This book examines the mind in action as it orchestrates skilled movement and how it understands the kinesthetic, symbolic language of dance.

There are new understandings of higher-level thought processes involved in learning, performing, creating, and appreciating dance in terms of what is now known about brain functions and skills necessary for lifelong success. All human movement initiates with thought, be it conscious or unconscious.[2] "The ability to think is cultivated in the process of developing the ability to move. The two grow symbiotically."[3]

Since the mid-nineteenth century, it has been known that separate areas of the brain provide different functions. The left and right hemispheres of the brain are not symmetrical and have neurons that are specialized. An understanding of brain locality has informed the field greatly. It gained popular recognition in the second half of the twentieth century when left brain/right brain information began to define talents and personality traits.

However, contemporary understanding of the brain through imaging demonstrates an orchestrated model. Brain areas function in symphonic harmony, which results in specific behaviors or emotions. Throughout the chapters of this book, specific localities are referenced, but each area contributes to an ultimate thought process and is not the sole perpetrator.

This book makes great effort to diminish the mind/body duality that splits contemporary understanding of the self. Yet there is a duality confronted by neuroscientists. While brain imaging has provided progress in understanding which areas of the brain function during specific behaviors and emotions, it cannot yet inform brain scan readers of a person's thoughts that produce the images, nor can it shed light on the life of experiences that culminate in those thoughts. It is a one-way street.

As medical imaging technology improved, the Dana Foundation[4] made neuroscience a primary priority. An Arts and Cognition Consortium was initiated in 2003 to study the effects of arts education on other learning domains. The Consortium included cognitive neuroscientists from across the United States,[5] and culminated with the publication of *Learning, Arts and the Brain*,[6] in which researchers grappled with the question: Are smart people drawn to the arts, or does arts training make people smarter? The findings for dance concluded that people who observe dance can learn dance.[7]

Neuroscientists and educators have collaborated in partnerships between major institutions to explore research on the brain, learning, and implications for education.[8] Most research in the arts applies to music and the visual arts. A few researchers have addressed movement, mainly about the effects of exercise on increased brain oxygenation. Exceptions are Steven Brown, who presents his work on neuroscience of the basic box step, and Emily Cross and Scott Grafton, who focus on observation.[9]

New information about learning and the brain is being produced in a variety of sources and spans different disciplines. Since there is little work examining brain functions in relation to dance, the purpose in this book is fivefold: (1) to introduce the thought process under discussion; (2) to explain pertinent brain processes; (3) to relate discoveries about learning and the brain to classroom practices; (4) to connect neuroscience discoveries and dance; and (5) to design movement explorations to help the reader experience content in the book.

The overall organization of the materials in this book progresses from simple theories involved with thinking and learning to those that are more complex. For example, observation and the ability to focus are of primary importance if learning is to occur. Yet these learning strategies are less complex in the educational continuum than nonverbal communication or the ability to connect one's feelings to learning experiences. Content also follows the logic of the creative process that develops from observation to synthesizing feelings, memories, and imagination to create movements and dances.

Chapter 1 lays the groundwork for the connection between brain functions, learning, and dance. The second chapter begins with visual observation and extends to other perceptual processes. In chapter 3, the focus is on engagement and focus as they relate to learning and dance. The next chapter includes information about higher-level thinking skills. Emotion, the topic emphasized in chapter 5, has an important function in all educational spheres because it promotes or inhibits learning. Chapter 6 approaches memory in its many forms, and chapter 7 investigates imagination. The final three chapters explore learning and problem solving, culminating with twenty-first-century skills.

We hope through this book you discover your own connections between neuroscience and dance, that they prove useful to you in your work, and provide insights with your students and other people in your life.

NOTES

1. Yvonne Rainer, "The Mind Is a Muscle," dance premiered in New York City at the Judson Church, 1966.

2. Mabel Todd, *The Thinking Body* (New York: Dance Horizons, 1937).

3. Rima Faber, "The Primary Movers: Kinesthetic Learning for Primary School Children" (master's thesis, American University, 1994).

4. The Dana Foundation is a private philanthropic organization established in 1950 by Charles A. Dana and is currently based in Washington, DC. It supports brain research through grants and educates the public about the successes and potential of brain research.

5. Dana Arts and Cognition Consortium, *Learning, Arts, and the Brain* (Washington, DC: Dana Foundation Press, 2008).

6. William Safire (1929–2009) chaired the Dana Foundation from 2000 until his death and focused the work of the institute toward an exploration of the neurology of learning and creativity and the contribution of the arts to both.

7. Dana Arts and Cognition Consortium, *Learning, Arts, and the Brain*.

8. The participating institutions were: Harvard University; Stanford University; University of California, Berkeley; University of California, Santa Barbara; University of Michigan; University of Oregon; and University of Toronto.

9. Stephen Brown, Michael J. Martinez, and Lawrence M. Parsons, *The Neural Basis of Human Dance* (Oxford: Oxford University Press, 2005); Learning & the Brain Conference, May 9, 2009, in Washington, DC; Scott Grafton and Emily Cross, "Dance and the Brain," in *Learning, Arts, and the Brain: The Dana Consortium Report on Arts and Cognition*, eds. Carolyn Asbury and Barbara Rich (New York/ Washington, DC: Dana Press, 2008), 61–69.

Acknowledgments

We wish to acknowledge the following individuals for their contributions to this book: Dr. Robert G. Shulman, Sterling Professor, Emeritus of Molecular Biophysics and Biochemistry at Yale University, for his generosity in writing the Foreword; Dr. Linda Lockwood, Experimental Psychology professor, Metropolitan State University of Denver, for reviewing the book and advising on the accuracy of our neurological analyses; and Sonya Everett for her illustration of the brain in figure 1.2.

Chapter 1

Groundwork for Thinking in Dance

Movement is the embodiment of thought; thinking directs action.[1] It has even been proposed that the main function of the brain is to create movement.[2] Survival in the animal world is dependent upon skillful and thoughtful action. A myriad of thought processes are involved in the learning and practice of dance. Dance is the language of physical movement. It communicates the values and beliefs of the society from which it emanates, and it projects meaning whether ritual, recreational, entertainment, or theatrical dance. Dance is a form of communication.

INTRODUCTION TO THE BRAIN

The adult human brain contains about one hundred billion neurons with trillions of synapses that are shaped by genetics.[3] The physical elements of the brain are ordinary: carbon, hydrogen, oxygen, sodium, nitrogen, phosphorus, iron, calcium, potassium, and a few others in trace amounts. The building blocks of the cells in a human brain are shared with other animal species: adenine, thymine, guanine, and cytosine. Some 98 percent of the DNA in human brain cells is the same as those in other mammals. Its uniqueness lies in its size and complexity, which makes it open to experience.[4]

The formation of the brain begins when an embryo is a few weeks old. A dense clump of cells forms at the top end of a tube, the precursor of the brain and spinal cord. They multiply rapidly, and one of each pair of cells migrates to the exterior layer and becomes differentiated for specialized roles and functions of the brain. The remaining partner stays in the primitive cortex to reproduce itself. In this way, neurons, dictated by DNA, change and spread throughout the brain and down the spinal cord into the embryo's body.[5]

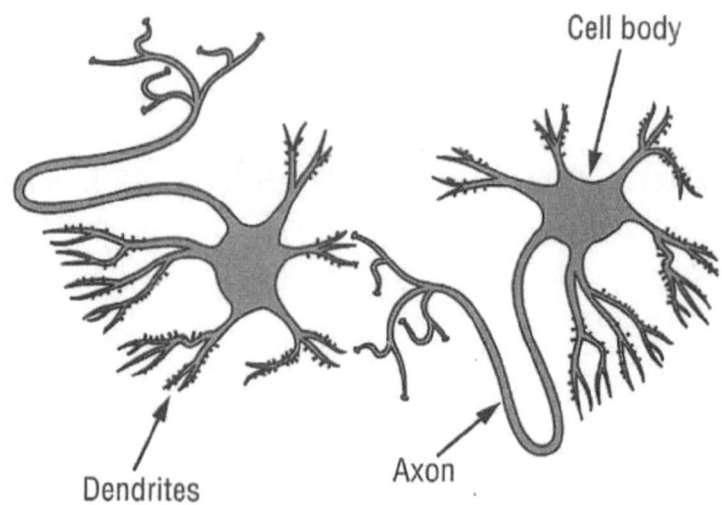

Figure 1.1 Typical neuron showing cell body, axon (output) and dendrite (input).
Sources: Teaching with the Brain in Mind, 2nd edition (figure 1.2, p. 9), by Eric Jensen, Alexandria, VA: ASCD. © 2005 by ASCD. Reprinted with permission. Learn more about ASCD at www.ascd.org.

Axons and dendrites in nerve cells fire to send and receive information through electrical impulses that travel from one nerve cell to the next. The flow is achieved due to chemical vehicles called neurotransmitters released at the end of each nerve fiber. The juncture that links each neuron to the next is called a synapse. A neuron that has branched synapses can transmit to as many neuron receptors as it has branches, or can converge impulses into as few as one neuron receptor. A neurotransmitter has to chemically fit the receptor "keyhole" for transmission to occur.[6] Keyholes cause neurons to engage in specific activities so that the entire system is not excited by each stimulus (see figure 1.1).

The concept of brain "localization" took hold during the mid-nineteenth century.[7] French neurologist Paul Broca explored the postmortem brain of his patient, Louis Leborgne, and found nerve lesions destroyed the left frontal lobe, producing an inability to speak or write, although he could understand speech and read.[8] This initiated the concept of brain localization in which sections of the brain control separate functions.[9] Carl Wernicke learned that damage to a posterior area of the left temporal lobe rendered persons unable to understand spoken or written language, although they could speak and read.[10] Modern technology has tempered the concept of localization. Although each area serves a function in thought, action, and sensation, they involve simultaneous orchestration of multiple brain areas. Figure 1.2 portrays important parts of the brain along with descriptions.

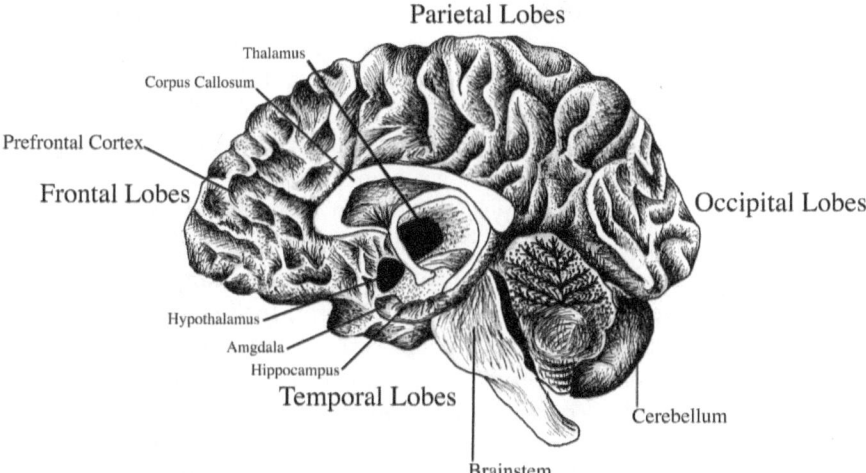

Figure 1.2 **Cerebral Cortex: Commonly called grey matter, it is the outer area of the brain and seat of awareness with convolutions or folds to provide greater surface area for greater neurological potential.** Impulses come to life to produce language, abstract thinking, spatial conceptualization, metacognitive perceptions and philosophical awareness. Corpus Callosum: It is a bridge between the brain's hemispheres and a highway through which orchestration is facilitated. Frontal Lobes: The right and left frontal lobes recognize experience differently, but coordinate to produce executive thinking. Propelled by the neurotransmitter dopamine, they are the seat of motivation, attention, planning, and short-term memory tasks. Parietal Lobes: Connections to our physical body are the domain of the parietal lobes. They perceive and integrate sensory input including proprioceptive and spatial awareness. Temporal Lobes: Different areas of the temporal lobes have very separate functions but, in general, are responsible for understanding or expression of language, retention or recall of memory, and processing sensory input into meaning or emotional association. Occipital Lobes: The occipital lobe is the visual processing area. It contains the visual cortex. Cerebellum: The cerebellum sits near the brain stem and receives input from the sensory and motor systems from the spinal cord and other brain areas. It does not initiate motor action, but is responsible for their coordination, fine tuning, accuracy, and precision of movement. Hippocampus: The hippocampi reside in the medial temporal lobes and are responsible for long-term memory. They are part of the limbic system and consolidate information from short-term to long-term memory. Amygdala: The amygdalae are two almond-shaped groups of nuclei, one in each hemisphere located deep in the temporal lobes. They play a primary role in processing emotional reactions, are part of the limbic system, and work closely with the hippocampi (memory) and the cortical system (decision-making). Thalamus: The thalamus sits centrally in the brain and serves to relay sensory and motor signals to the cerebral cortex. It regulates consciousness, sleep, and alertness. Hypothalamus: It is part of the limbic system that provides a link from the nervous system to the endocrine system via the pituitary gland, regulating the production of hormones. Brainstem: The spine connects into the posterior part of the brain as a continuous mechanism that brings bodily sensory and motor information into the brain. *Source*: Sonya Everett © 2015.

Chapter 1

THE MIND, BRAIN, AND BODY CONNECTION

As a result of recent investigations in neuroscience, it is now recognized that the basis of thinking and bodily movement is a result of intricate electromagnetic, chemical, and physical processes that occur simultaneously throughout the brain. Thoughts and messages are transmitted by neural connections through the body to relay the physical nature of thought and mind.

Philosophers, researchers, educators, and dance theorists, past and present, have acknowledged the physical basis of thought and learning. Nerves do not think by themselves. In the theory of felt-thought, feeling and meaning are physical and intimately connected. There is a general reorganization in the physical body that comes before one is able to clearly conceptualize or define a thought.[11] The whole person imagines and creates.

The study of brain functions in relation to language has unearthed valuable knowledge. Research performed on stroke patients or victims of brain injury have provided much information about different areas of the brain and their impact on both motor and thought processes. The frontal left temporal lobe bears Broca's name, and dysfunction produces a condition called expressive aphasia, the inability to communicate verbally. Damage to Broca's area does not impact memory of words but the function of symbolic abstraction in language, which prevents recall involving abstract thought (see figure 1.3).

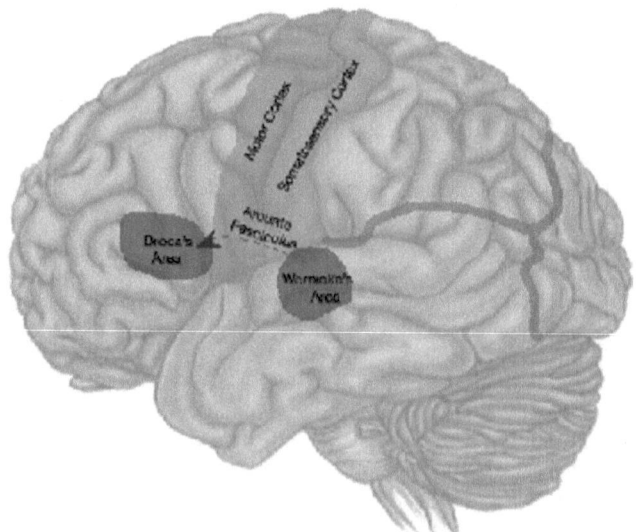

Figure 1.3 Broca's and Wernicke's areas, Motor and Somatosensory Cortices.
Source: From *The Tell-Tale Brain: A Neuroscientist's Quest for What Makes Us Human* by V. S. Ramachandran. Copyright © 2011 by V. S. Ramachandran. Used by permission of W. W. Norton & Company, Inc.

In 1976, a woman we shall call Melissa suffered a Broca stroke. Within ten minutes of initial symptoms, she could not speak or write, but emitted guttural mumbling. After a week Melissa spoke a "word salad," a verbal jumble that was actual words, but not appropriate to her meaning. One morning Melissa anxiously repeated, "Superintendent, superintendent!!" This was associated to her apartment. Her dog was in her apartment, so she was reassured that the dog was being attended to by her neighbor, and she relaxed with a smile.

Because Broca's area is one of the major areas responsible for expressive language, Melissa could read and understand language perfectly, but she could not recount what she had just read. Having lost the ability to use abstract symbolic language, speech was concretely connected to actions. Each morning her doctor would walk into the room, hold up his pen, and ask, "What is this?" Melissa could not tell him. The name given to an object is an abstract symbol until the name is related to its function or use. Melissa could name the pen when she wanted to write, or ask for soap when she wanted to wash, but not when words were disassociated from their functional activity.

Since the left hemisphere of Melissa's brain was severely damaged, she was also completely paralyzed on her right side. As Melissa's mind struggled to make her muscles move again, her mental exertion was, literally, painful. She was in fabulous physical condition when she suffered the stroke and was running five to ten miles daily. Unfortunately, within twenty-four hours without any neurological stimuli, muscles lose 99 percent of their strength. In addition, when nerves are destroyed or die, they do not heal. Neural connections must be newly created.

Recovery began with mental imagery. The physical therapist would move one of her body parts involving a muscle group while Melissa concentrated on the motion. Gradually, new neurological connections were developed to achieve gross motor mobility, but it required enormous concentration. Basic movement skills—for instance, walking—require fine motor actions that are not automatic, and finer skills require additional focus.

Even after one month in the hospital and three months of allotted home therapy were completed, Melissa's quality of motion was still spastic, jerky, and her speech was primitive. Moshe Feldenkrais's somatic system of minimal movement to repattern deep neurological habits or issues. It proved highly effective to enable Melissa to regain her fine motor skills.

SOMATIC PRACTICE

Throughout the past century, there have been systems of movement that use the mind/body to teach fluid or expressive movement. Inspired by Isadora Duncan, Florence Noyes used imagery from nature to help women overcome movement inhibitions and self-consciousness about their bodies.[12] In 1937, Mabel Todd intertwined thought and movement as simultaneous action. Lulu Sweigard later developed Todd's work as ideokinesis.[13] The work of Rudolf Laban became widely recognized as Laban Movement Analysis, which spawned Irmgard Bartenieff's system of Bartenieff Fundamentals. These somatic systems extend movement possibilities.

The body as a basis of knowing has been an integral part of somatics practices, which have entered dance training to release tension, produce movement free of stress, and provide more efficient actions. Practitioners integrate body and mind to help their clients transform themselves through increased body awareness in their movement. Somatics unites the body/mind split to achieve optimal function.

The Alexander Technique, created by Frederick Matthias Alexander, is a widely known somatics system particularly helpful to musicians, actors, and dancers. By analyzing his personal difficulties with vocal projection as an actor, Alexander learned to "recognize and overcome reactive, habitual limitations in movement and thinking."[14] The client is coached in the correct use of the body, especially functional alignment of the head in relation to the spine, as a way of creating ease in movement and ridding the body of discomforts.[15]

Moshe Feldenkrais developed a system of private instruction called Functional Integration and class instruction called Awareness through Movement.[16] Conscious integration of body and mind facilitates deep neurological patterning of movement to rid the body of inefficient movement patterns through the coordinated use of the mind to produce physical changes. The minimal movement exercises invented by Feldenkrais develop fine synchronous neurological patterns that are consciously produced until they become automatic habits on a subconscious level.

Bonnie Bainbridge Cohen developed Body-Mind Centering, which is based on gaining bodily wisdom through awareness. As her students explore expression through movement they become aware of movement habits and qualities, and learn how the mind moves or restricts the body. Body-Mind Centering is an experience of developmental movement patterns using conscious repatterning.[17] Cohen related the evolutionary movement of animals to the developmental movement of humans beginning with the generalized motions of newborns prevertebrate wriggling, reptilian slithering, crawling, upright sitting, and ultimately primate walking.[18]

Movement Explorations

Alexander Technique: (a) Walk around a room or open space and be aware of how you usually walk. Where is the weight on your feet? How do you use your legs? Your hips? Your arms? Your shoulders? Your neck? (b) Apply the following Alexander principle to the way you walk. Feel your head traveling gently upward (from your cowlick) and slightly forward. Feel the back of your neck lengthen, and feel like your head is leading your body upward and slightly forward. (c) How does this affect your walking and each part of your body as you walk?

Feldenkrais Method: (a) Sit on a chair. (b) Rise to standing, and sit back down. (c) Apply the principle that "performance is improved by the separation of the aim from the means."[19] To do this, repeat the action, but instead of thinking of getting up, think about raising your seat and at the same time get up. Instead of thinking about sitting down, lower your seat back into the chair and complete the movement with this in your mind. (d) Reflect on the different sensation of these actions.

Bartenieff Fundamentals: This exercise strengthens the diagonal connections between your upper and lower body. (a) Lie on the floor with your knees bent and your feet flat on the floor. (b) While keeping your pelvis anchored to the floor, gently drop your knees to the right side. (c) After pausing, extend your left arm diagonally upward from your shoulder and trace a large, counterclockwise circle with your left arm overhead, across your body and down to the floor, following your arm with your eyes. (d) Pause and reverse the direction of your arm circle. (e) What changes did you feel in your body? (f) Repeat this exercise on the other side.[20]

Body-Mind Centering: This somatic system explores the internal body through movement. The following focuses on the lungs, but can apply to any organ. (a) Breathe (inhale and exhale) into the lungs. What does your breath feel like inside your lungs? What effect does inhalation have on your body? Your exhalation? (b) Exhale with a hissing sound. What effect does your sound have? (c) Explore sounds with variations in pitch and intensity. How do they affect the lungs? (d) Initiate movement from the lungs. What movements arise from them? (e) Reflect on any insights that have emerged.[21] (f) Extend this exercise by applying explorations a–d to other organs; for example, heart and brain.

LABAN MOVEMENT ANALYSIS

Rudolf Laban was a noted dance artist and theorist who developed a body of work adopted by artists and scholars in Europe and the Americas. He created an integrated system for observing, describing, and notating expressive movement that became standard for teaching and learning dance. Laban's categories describe movement patterns that define the basic elements of dance as space, time, and force, and qualities of movements as "efforts."[22] Meaning in movement is communicated through relationships of the elements and efforts.

Space includes the three-dimensional characteristics of environment and the body; the relationships of facings, pathways, levels, shapes, design, the size or range of movements, direction, and whether they extend outward away from the body or inward toward the center. The use of space can be direct or indirect. All dance involves the human body existing in space, even if the dancer is stationary.

Time refers to duration. It is defined by tempo, beat, rhythm, meter, and accent. Humans pattern time and create rhythms that organize and define time, most often recognized in sound. Bodily motions as basic as walking establish movement rhythms. Laban defines the polarities of time as quick or sustained. Dance is greatly affected by rhythmic patterns, as evidenced by rhythmic differences between tap and ballet.

The term *effort* currently has many names: force, dynamics, or energy. These involve weight (strong/light), flow (bound/free), attack (sharp/smooth), and quality (sustain, suspend, collapse, percussive, swing, vibratory). Laban defines effort qualities as: press, punch, dab, flick, slash, wring, float, and glide, all of which relate to space, time, weight, and flow. Laban's system became codified as Laban Movement Analysis (LMA), and the written symbols are known as Labanotation.

Movement Explorations

Space: (a) Think of a simple action that travels through space (walk or run). (b) Choose a direction through which you will travel. (c) Perform your action in a specific pathway (straight line, curved, or zigzag). (d) Perform your action again changing levels (low, middle, high). (e) Finish your action by making a shape in place with your body (bent, straight, or curved).

Time: (a) Clap a steady beat at a comfortable tempo. (b) Stamp your feet or tap your toes on every other beat. (c) Stamp or tap your toes twice as fast as your clapped tempo. (d) Reflect on how the different tempi feel in your body.

Force/Effort: (a) Return to the simple action you chose for the first exploration. (b) Weight—Imagine you are performing your action on Jupiter (very heavy). Then, perform your action as if you were on the moon (buoyantly light). (c) Flow—Perform your action as if you are pushing a rock. Then, perform your action as if you are floating in the wind. (d) Attack—Imagine you are performing your action as if you are slashing through a jungle with a machete. Then, perform your action as if you are swimming in a smooth, clear lake. (e) Reflect on how your body felt performing the different movements.

BRAIN-COMPATIBLE MOVEMENT PATTERNS

Anne Green Gilbert developed Laban/Bartenieff's movement patterns into a progression relating to brain function. She outlined eight basic movement patterns that exercise the brain: (1) full, deep breathing; (2) tactile stimulation of the skin; (3) core-distal actions alternating direction toward the body and reaching out; (4) head-tail movements that involve moving the head toward the coccyx; (5) upper-lower body organization that mobilizes the upper while stabilizing the lower half or vice versa; (6) body-side where only one side of the body is moved; (7) cross-lateral involving moving across the body's midline; and (8) vestibular activation by spinning in alternate directions.[23]

Movement Explorations

Breath: Breathing oxygenates the brain and body. It is the newborn's initiation to the world. (a) Place your hands on your ribs and inhale as deeply as you can. Do you feel your ribs expand? Exhale and feel your ribs as they collapse when your diaphragm returns to resting position. (b) Repeat this cycle slowly five times. (Stop immediately if you feel dizzy or light headed.)

Tactile: Skin covers the entire body, includes a vast system of nerves, and produces a myriad of sensations. (a) Pat yourself all over your arms, legs, and torso. (b) Poke (dab) yourself. (c) Rub yourself vigorously. (d) Smoothly stroke yourself. (e) How does your skin feel? How does your energy level feel?

Core-Distal: Movement close to or toward the body engages the core muscles versus movement that reaches out into space. (a) Sit on the floor or in a chair and curl up tight (small) as you can by engaging your core muscles. How does your body feel? (b) Open your body, leading with your arms and legs (extremities) and reach out into space as widely as you can. How does your body feel? (c) Alternate ten times from one to the other. (d) Does this affect your sense of body in space?

Head-Tail: Nerves from the brain extend down the spinal column and branch out throughout the torso, limbs, and organs. (a) Sit on the floor or in a chair and curve your spine forward so your head travels toward your legs. Picture the top of your spine circling toward your coccyx. (b) Reverse the direction of your head and arch backward. Picture the top of your spine circling back toward your coccyx. (c) Bend or twist sideways, feeling your head and hip moving toward one another. (d) Change from one to the other five to ten times. (e) Reflect on how your body feels.

Upper-Lower: Mobilize the upper body while stabilizing the lower half, and vice versa. (a) Stand in one place and move your upper body. What is your sense of the movement? (b) Keep your upper body completely still and move only your lower body. What is your sense of the movement? (c) Alternate upper and lower body movements ten times. (d) Reflect on how your body feels.

Body-Side: The left side of the brain moves the right side of the body, and activating the right hemisphere moves the left side of the body. Even so, same-side movements are less complex than movements crossing the midline. (a) Move your right leg and right arm (sequentially or simultaneously) without crossing the midline of your body. (b) Move your left leg and arm (sequentially or simultaneously) without crossing the midline of your body. (c) Alternate ten times from one to the other. (d) Reflect on how difficult this was to coordinate and the thought process required.

Cross-Lateral: Movements involving diagonal body quadrants require simultaneous use of both brain hemispheres. (a) Standing or sitting, execute movements of the right arm or leg that cross the body midline to interact with the left side of the body. (b) Reverse the movements by using the left arm or leg to cross the midline toward the right side of the body. (c) Cross quadrants (upper-right quadrant interacts with lower-left quadrant and vice versa), connecting with movement in many ways. (d) Alternate five to ten times, initiating on one side and then the other. (e) Reflect on how difficult this was to coordinate and the thought process required.

Vestibular: Spinning massages the brain with a bath of synovial fluid. (a) Spin in place for ten to fifteen seconds in one direction. (b) Remain still for fifteen seconds. (c) Spin in place for ten to fifteen seconds in the opposite direction. (d) If you are dizzy, focus on a hand or finger in front of your eyes. (e) Reflect on how your body and mind feel from doing this series of exercises.

DANCE AND COGNITION

Cognition pertains to how we think to attain knowledge and understand the world. Cognitive sciences gained recognition in the last century with work of Lev Vygotsky, Alexander Luria, Heinz Werner, and Jean Piaget. Before the work of these pioneers, it was assumed thought processes of children were the same as those in adults, but with less life experience and information at their disposal. Research demonstrated that children perceive the world in a qualitatively different perspective that shapes their understanding of reality.

Piaget outlined four phases in cognitive development. During the Sensori-Motor stage, ages zero to two, an infant's perceptions are based on sensate awareness and bodily movements. They are: egocentric, a view of the world in which they are the center of perspective and causality. Children ages two to seven function in a Concrete Pre-Operational stage. Although sense of the world is still interpreted concretely, enormous neuron development occurs, leading to language and development of the corpus callosum. Children at the Concrete Operational stage (ages seven to eleven) are literal minded, grounded in how things work, and understand the world in absolute values. During the Formal Operations stage, ages eleven to sixteen, teens become wired for metacognition and abstract thinking.[24]

Piaget's delineation of stages has fallen in and out of favor in cycles, but it is mentioned here because many of their attributes are explored in later chapters. A developmental understanding provides insight into the thought and learning processes, especially for teachers and dance educators.

THE MANY FACES OF INTELLIGENCE

Domains of learning have been recognized since the 1970s, but in 1983 Howard Gardner introduced a new concept of intelligence, a Theory of Multiple Intelligences (MI). Gardner defines intelligence as the ability to problem solve in the practical world as opposed to a score on the Stanford-Binet paper-pencil test. He proposed first seven, and then an eighth, intelligence: linguistic, musical, mathematical-logical, spatial, bodily-kinesthetic, intrapersonal, interpersonal, and natural.[25] Given the demands of dance on the body/mind, dance develops each of them.

Bodily-Kinesthetic Intelligence: "the ability to use one's body in highly differentiated and skilled ways."[26] It involves gross and fine motor coordination in service of cognitive solutions to physical goals, including the aesthetic of dance.

Linguistic Intelligence: verbal abilities, both written and spoken. Although dance is the language of movement, dancers and dance educators need verbal skills to explain the intent of movement or a dance work to a cast or students. In addition, dance often uses literary themes, narrative structure, or spoken text. Choreographers, especially, are often linguistically gifted.

Musical Intelligence: musical sensitivity and comprehension. The body is the dancer's artistic instrument. Historically, dance has been performed mainly to musical or rhythmic accompaniment, and often reflects the musical intent and structure. Although contemporary dance forms are sometimes accompanied by text, sounds, or silence, the dancer must perform with clear aesthetic movement phrasing.

Mathematical/Logical Intelligence: facility with the language and logic of math as a communication of relationships. Dance, like physics, is the exploration of moving bodies in time and space, and the physical properties and relationships of the human body. To choreograph for a group of dancers requires mathematical juxtapositions and logic that results in a progression

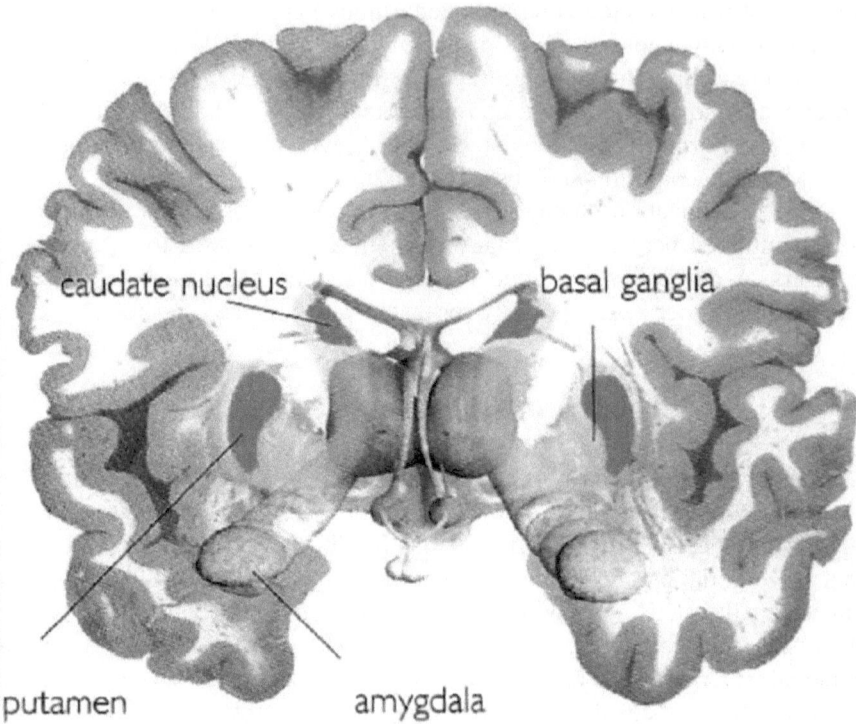

Figure 1.4 Basil ganglia and other deep parts of the brain. *Source: Mapping the Mind,* by Rita Carter. © 2010 by Rita Carter. Published by the University of California Press.

of patterns and phrasing in a dance. Dancers learn movements to counts, which means dance movements are measured in time and remembered in sequence.

Spatial Intelligence: equated with the visual arts, but who has more spatial intelligence than the blind? Dance develops a visceral sense of space; a keen awareness of direction, orientation, spatial composition, and adaptability to changing spatial environments. Space is one of the basic elements of dance.

Intrapersonal Intelligence: insight into oneself and sensitivity to one's own feelings. Dance is demanding in that one has to surpass bodily limitations. Confrontation with those limitations requires self-improvement, which sometimes necessitates overcoming a great many personal hurdles.

Interpersonal Intelligence: communicative and social abilities. Dance is a social art form involving ensemble cooperation, coordination, and sometimes collaboration. It requires great skill and awareness working with colleagues, choreographers, and audience populations. Dancers often teach for financial sustenance, which requires sensitivity and insight into students' needs.

Naturalist Intelligence: understanding and appreciation of nature. It is believed prehistoric dance began as spiritual expression; as prayer for needs, for survival, communion with nature, or to offer thanks to the gods for blessings. Although most performance-based dance forms have become removed from these tribal characteristics, there is need for sensitivity to the body in relation to the ground, gravity, and natural forces.

These are holistic neurological processes. To achieve bodily-kinesthetic movement, large parts of the cerebral cortex, the thalamus, basal ganglia, and cerebellum feed input to the spinal cord. This means the movement system is very complex[27] (see figure 1.4).

CREATIVITY

Philosophers and later psychologists and neurologists have sought to discover the source and ingredients of creativity. As the millennium turned its corner, Robert and Michèle Root-Bernstein examined the type of thinking used by the world's most creative people from many disciplines. Their research resulted in a categorization of thirteen thinking tools used in the creative process. The thinking tools are: observing, imaging, abstracting, recognizing patterns, forming patterns, analogizing, body thinking, empathizing, dimensional thinking, modeling, playing, transforming, and synthesizing.[28]

Observation: to observe movements clearly and in detail; a process of perception, that according to the Root-Bernsteins is primary in creative thinking and problem solving. Perception includes reception and awareness from all the senses with notice of movement details such as lines, shapes, textures, patterns, and sounds accompanying a dance. Creating movement compels an observation of the inspiration or motivation.

Imaging: a process of imagining or recreating experience in the mind and expressing it through the body. "As Martha Graham makes clear, most imagining is actually polysensual."[29] Graham taught and choreographed through powerful imagery. In one of her classes, Graham wanted students to perform a movement as if holding and rocking a baby. She demonstrated a movement from *Appalachian Spring* in which the image was so visceral she seemed to be actually holding a baby. Movement can be both imagined and felt.

Abstraction: the essence of a subject that pares away extraneous elements. The process of creating art in any discipline, including dance, is a process of selection, or abstraction from all possibilities of experience or elements of artistic expression to focus on those that communicate the artist's intent.

Recognizing Patterns: The human brain organizes experiences into patterns, one of its main capabilities that separates human cognition from brain functions found in animals. Language could not occur without the ability to recognize patterns of sound.

Forming Patterns: creating relationships between discrete entities that are part of a culturally developed aesthetic. The choreographic process involves creating movement sequences or patterns that build to form a dance.

Analogy: a form of comparison between things that have similarities. Doris Humphrey created movement orchestrations of Bach concertos. José Limón choreographed his version of Shakespeare's *Othello*. These choreographic works are analogous to the original music or play.

Body Thinking: sensations when performing overtly physical actions. "Thinking with the body depends on our sense of muscle movement, posture, balance, and touch."[30] The mind/body connection heightens awareness of proprioceptive experiences of bodily movement.

Empathizing: a reflective response to incoming stimuli producing vicarious experience or thoughts of another individual. The Root-Bernsteins published their work before there was an in-depth understanding of mirror neurons and the operation of empathy. Yet they reported, "Isadora Duncan understood that dance, like music, must stimulate empathy within the bodies of onlookers, creating within them the desire to move."[31]

Dimensional Thinking: a conceptual, artistic, and visceral understanding of space; perceptual awareness of the direction, shape, and patterns of movement. Dancers make both me-maps and mind maps of their movements in space and time.[32]

Modeling: the production of larger-than-actual or miniature replicas of creative work in fields such as architecture, sculpture, or science. Learning, especially in dance, involves a process of internalizing a model of what is to be learned. This is often achieved through demonstration by the teacher or from observation of an excellent performance.

Playing: a form of improvisation. In dance, it involves inventing movement by exploring an idea, space, or objects to discover new movement possibilities. The main difference between exploration, experimentation, and play is the goal, and the use of outcomes. "Playing is therefore more than just exercising other tools for thinking; it is a tool in and of itself."[33]

Transformation: "Take a look at any creative endeavor and it is possible to find ideas and insights transformed through many tools for thinking and translated into one or more expressive languages." In addition "different transformations of an idea or set of data will have different characteristics and uses."[34] An example is the transformation of movement into different types of notation. Likewise, choreography is the transformation of an idea for a dance into movement.

Synthesis: the end result of transformational thinking in which impressions, feelings, knowledge, and memories come together in a unified way.[35] Synthesizing is the summation and culmination of all the thinking tools. It joins sensory, experiential, and intellectual processes into the creative moment.

It results in creative fullness, when you know a work of art is complete. A completed dance work represents the achievement of a synthesis.

Movement Explorations

Observing: (a) Observe an object. (b) Explore the qualities of the object through movement.

Imaging: (a) Picture a fairly simple image in your mind, and either move like the image or make the shape of the image with your body. (b) Reflect on how your body relates to the image.

Abstracting: (a) Return to your previous image, or change it if you wish. (b) Choose one aspect of the image. (c) Ask yourself, What is the basic essence of this aspect of the image? (d) Create a movement or shape based on that essence.

Recognizing patterns: (a) Reflect on the movement or shape you developed for your image. (b) Describe some of the patterns you observed in your movement or shape.

Forming patterns: (a) Reflect on the patterns observed above. (b) Develop these patterns further in movement to expand on ideas relating to your image. (c) How would you describe the movement patterns?

Analogizing: (a) Reflect on how your image can relate to other areas of life or artistic meaning. (b) Develop the movement you have created to include these wider perspectives. (c) Reflect on how analogizing expands choreographic and creative potential.

Body thinking: (a) Think about the movements or gestures you have created. (b) Focus on the body sensations, feelings, or meanings they call forth.

Empathizing: (a) Focus on the movements you have created. (b) Think about how your movements would communicate to another person. (c) What would someone watching you feel or experience from your movements?

Dimensional thinking: (a) Expand your movements into space using a variety of: directions, sizes, levels, pathways, shapes, and relationships. (b) Reflect on how moving in the different dimensions changes your choreography. (c) What effect does dimensionality have on movement?

Modeling: (a) Using the movements you have developed, work with a partner and instruct the partner only verbally while teaching your movement. Do not show them anything physically. (b) Reverse the process so that your partner teaches their movement only physically without talking. (c) Reflect with your partner on how the two learning experiences differ. Discuss how modeling movement affects your learning process in comparison to learning a movement from a verbal description.

Playing: (a) Improvise using ideas inherent in your image and ideas stimulated from your choreography. (b) If you have done the previous exploration with a partner, you and your partner could improvise together and spark ideas from one another. (c) Experiment with the movements you have created to see how many different ways they can be performed. (d) Reflect on how this process develops your choreography.

Transforming: (a) Reflect on how the preceding steps have transformed your original image into movement. (b) Reflect on how the preceding steps have transformed the meaning of your original image.

Synthesizing: (a) Select movements developed from the first twelve tools that best fulfill the artistic meaning and arrange them into a movement sequence. (b) Mesh the performance of your choreography with its meaning to deepen the movement and give it significance. (c) Show it to at least one other person, and discuss how the thinking tools of dance were applied to the creative process.

LEARNING AND MOVEMENT

In recent years, there has been a growing interest in kinesthetic learning or dance integration—the use of movement and the body to teach concepts from other academic disciplines. Bodily action produces neural growth and greater neural density. The cerebellum, which is a part of the brain traditionally thought to be important in maintaining posture, balance, and movement, also has a role in coordinating brain functions of separate areas of the brain related to learning.

Movement can make learning efficient. The brain is most stimulated by and attracted to change. The use of movement as a teaching strategy is a novelty for students that engages the brain and captures student attention. Use of movement demands and creates energy and therefore helps students pay attention and maintain interest. The brain operates most immediately on concrete experiences, and movement-based lessons make content more concrete and hence less complex to learn. Movement in the classroom is enjoyable, and information is easier to learn and remember when it is connected to a positive emotional state.[36]

Dance integration practices assist in understanding concepts from a variety of subjects and promote successful student achievement. Teachers who used integration strategies reported success with their students.[37]

Movement Explorations

Language Arts: (a) Select a short story. (b) Think of one image that represents the beginning of the story and another than represents the end. Create a body shape that communicates each image. (c) Analyze the story to determine its main idea, and develop a body shape that communicates it. (d) Start in the beginning shape, create movements that lead into the main idea shape, and movements that develop to the end shape. (e) Perform your dance based on the story. Did creating movements give you a different perspective on the story and its structure?

Math—Fractions: (a) In a large open space, start at one end, and mark your starting point with a line. (b) Count with your feet (heel to toe) how many "feet" the whole space measures, and mark the end point. (c) Perform a movement across the full space (e.g., run, jump, hop). How many kinds of movements were done? (one) (d) Divide the whole by two, and count that many feet to mark the midpoint, which is the half point. (e) Start at one end, jump to the half point, and turn to the end. How many types of movements have you done? (two) How many halves make a whole? (two) (f) Continue to subdivide the space as many times as desired and perform different movements for each fraction of the whole space.

Science—Photosynthesis: Photosynthesis is the process a plant uses to eat and breathe for energy. It ingests water (H_2O) and carbon dioxide (CO_2), absorbs sunlight (energy/heat) to act as a catalyst with the CO_2, carbon (C), and chlorophyll in the leaves of the plant, and releases oxygen (O). (a) Use movement to explore the qualities of each of the elements and how they function in photosynthesis: H_2O (sprinkled, fluid, or flowing); CO_2 (dense, heavy gas); sunlight (bright, active, and jumpy); C (hard, thick, or stiff); chlorophyll (green leaves); and O (light, floaty gas). (c) Dance the process in the order in which it occurs in nature.[38]

Social Studies—Geography and Multiculturalism: (a) Choose two countries that have different types of typography, such as urban, forested, mountainous, and desert. (b) Learn at least part of a dance from each country. (c) How do these movements reflect the environment of each culture?

NOTES

1. Jay Seitz, "The Bodily Basis of Thought," *New Ideas in Psychology*, Elsevier Science Lt., http://postcog.ucd.ie/files/the_bodily_basis_of_thought%20%20Seitz%202000.pdf; Mabel Todd, *The Thinking Body* (Brooklyn: Paul B. Hoeber, 1937).

2. Daniel Wolpert, "The Real Reasons for Brains," accessed July, 2014, http://www.youtube.com/watch?v=7s0CpRfyYp8.

3. Gary Marcus, *The Birth of the Mind* (New York: Basic Books, Perseus Books Group, 2004).

4. Ann Barnet and Richard Barnet, *The Youngest Minds* (New York: Simon & Schuster, 1998).

5. Barnet and Barnet, *The Youngest Minds*.

6. Barnet and Barnet, *The Youngest Minds*.

7. Brain localization refers to the fact that human functioning occurred in discreet areas of the brain.

8. Pierre Paul Broca (1824–1880) was a French neuroanatomist whose research on the localization of speech led to entirely new research into the lateralization of brain function.

9. Michael Ruane, "Brain Teaser," *Smithsonian Magazine* (August 2013): 20.

10. Wernicke's area was first described by a German neurologist, Carl Wernicke, in 1874.

11. Harold Rugg, *Imagination* (New York: Harper & Row, 1963).

12. Rima Faber, "The Primary Movers: Kinesthetic Learning for Primary School Children" (MA thesis, American University, 1994).

13. Mabel Todd, *The Thinking Body* (Brooklyn: Dance Horizons, 1975); Lulu Sweigard, *Human Movement Potential: Its Ideokinetic Facilitation* (New York: Dodd, Mead & Co., 1974).

14. "Alexander Technique," Wikipedia, accessed July 2014, http://en.wikipedia.org/wiki/Alexander_technique.

15. Mirka Knaster, *Discovering the Body's Wisdom* (New York: Bantam Books, 1996).

16. Moshe Feldenkrais, *Awareness through Movement* (New York: Harper & Row, 1972).

17. Linda Hartley, *Wisdom of the Body Moving: An Introduction to Body-Mind Centering* (Berkeley: North Atlantic Books, 1995).

18. Bonnie Bainbridge Cohen, *Sensing, Feeling, and Action* (Northampton, MA: Contact Editions, 1993).

19. Feldenkrais, *Awareness through Movement*, 81–82.

20. Knaster, *Discovering the Body's Wisdom*.

21. Bainbridge Cohen, *Sensing, Feeling, and Action*.

22. Rudolf Laban, *Modern Educational Dance* (London: MacDonald & Evans, 1968).

23. Anne Green Gilbert, *Brain-Compatible Dance Education* (Reston, VA: National Dance Association, 2006).

24. Jean Piaget, *The Child's Conception of the World* (Patterson, NJ.: Littlefield, Adams & Company, 1963).

25. Howard Gardner, *Frames of Mind* (New York: Basic Books, 1983).

26. Gardner, *Frames of Mind*, 206.

27. Gardner, *Frames of Mind*, 210.

28. Robert and Michèle Root-Bernstein, *Sparks of Genius* (New York: Houghton Mifflin, 1999).

29. Root-Bernstein and Root-Bernstein, *Sparks of Genius*, 59.

30. Root-Bernstein and Root-Bernstein, *Sparks of Genius*, 161.

31. Root-Bernstein and Root-Bernstein, *Sparks of Genius*, 183.

32. Root-Bernstein and Root-Bernstein, *Sparks of Genius*.

33. Root-Bernstein and Root-Bernstein, *Sparks of Genius*, 248–49.

34. Root-Bernstein and Root-Bernstein, *Sparks of Genius*, 285.

35. Root-Bernstein and Root-Bernstein, *Sparks of Genius*, 296.

36. Traci Lengel and Mike Kuczala, *The Kinesthetic Classroom: Teaching and Learning through Movement* (Thousand Oaks, CA: Corwin, 2010).

37. Sandra Minton, *Using Movement to Teach Academics: The Mind & Body as One Entity* (Lanham, MD: Rowman & Littlefield Education, 2008).

38. Rima Faber, "Science with Dance in Mind" (final report, Baltimore County, October 2011).

Chapter 2

Observation

We encounter a multitude of observations and sensations during the course of a minute. The goal of this chapter is to focus on observation as a primary sensory input used in the beginning stages of critical thinking, creative problem solving, and learning.

Although observation is noted with all the senses, sight is generally considered the most prevalent tool of perception. Vision begins with the stimuli of light passing through our eyes. Sight becomes recognized, understood, and artistically appreciated through a complex thought process. Vision does not occur in the eye. It occurs in the brain. The human visual system results in vision that is different from any other animal due to the way our brains interact with sight.

THE HUMAN OPTIC SYSTEM

The eye provides a pathway for light images to enter the brain. Vision begins as an image enters the pupil, a dark cylindrical opening in the eye, and travels through the iris, lens, and cornea. The lens regulates the light's intensity by contracting when bright and widening when dim. The curved cornea serves to bend light into focus where the image enters the internal eyeball and converges to a focus in the center. If focus is too close to the front of the eyeball, one sees close objects clearly, but distant vision is blurry. The reverse occurs when the focal point is behind the center of the eye. At the focal point the image flips upside down, reaching the retina, and entering the optic nerve inverted. The brain re-reverses it, so we experience the images in correct orientation.

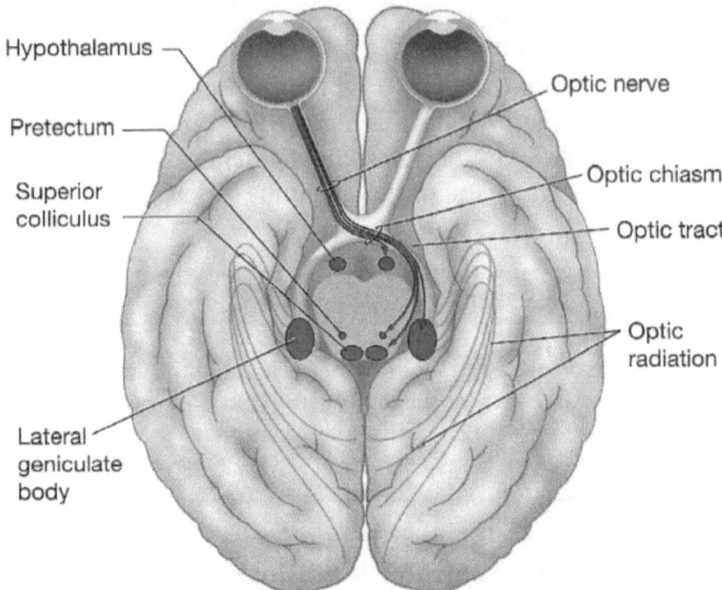

Figure 2.1 Superior colliculus and other optic tract projections. *Source*: Schenkman, Margaret; Bowman, James; Gisbert, Robyn; Butler, Russell, *Clinical Neuroscience for Rehabilitation*, 1st © 2013. Reproduced by permission of Pearson Education, Inc., New York, New York.

Brain Network

Light-sensitive cells in the retina turn light into electrical impulses.[1] Electrical impulses are carried along the optic nerve from each eye, and part of the information received at the retina crosses over at the optic chiasm. The optic tract then carries visual information from each visual field to the opposite side of the brain (see figure 2.1). The occipital lobes are considered the centers for sight, but parts of the temporal and parietal lobes are also actively devoted and contain a complete or partial map of what one sees. However, brain maps are recreated as neural impulses, not in the form of visual images.[2]

The visual cortex has specialized areas for processing different aspects of sight, such as color, shape, and size. When a person looks at an object, a pattern of neuronal activity is created on the surface of the visual cortex that matches a stimulus in the visual field.[3] The visual cortex also returns images to their proper right-side-up and right-left orientation.[4]

The Nobel neuroscientist Eric Kandel noted that visual sensations are first deconstructed into submodalities (spots of light, contours, colors) and automatically reconstructed in the cortex. Specialized cortical cells readily respond to linear contours or edges between dark and light similar to outlines of objects in the environment, and to angles and edges that are either vertical or horizontal.[5]

Classroom Applications

Observation is pervasive in learning for individuals with sight. Some students, however, have better visual organization than others, and are stronger visual learners. They process information globally and use mental imagery or mind pictures to aid understanding. Visual learners can learn to read from the shapes of letters. They prefer to look at information in the form of charts, drawings, diagrams, or outlines. Presenting visual information in a way that copies how the visual cortex processes information aids learning. For example, the teacher could emphasize the color, size, or shapes represented in visual input to help students learn information.

Movement and Dance Connections

Dancers learn most movements through observation of a physical demonstration. Aside from physical demonstration, observing video is currently a common system of passing on repertory. Early investigations suggest observation of movement provides a blueprint for actions to be learned, followed by practice. Some think this perceptual blueprint is a means to detect and correct errors.[6] Other findings indicate the fastest and most accurate way to learn movement is through observation with simultaneous physical practice.[7]

Researchers have uncovered a phenomenon caused by mirror neurons. When a person observes something, their brain resonates with reciprocal impulses or patterns that mirror the incoming stimuli. The result is an internal neurological duplication of the external experience called empathy. This produces learning from whatever is observed.[8]

The eye encodes neurological impulses for contours, angles, shapes, lines, and the ever-changing designs of the dancing body. It is attracted to novel patterns and colors. A dance leads the audience's eyes through a journey, and empathy internalizes the experience.

Movement Explorations

Scanning the Visual Field: (a) Find a YouTube video of choreography created by Alwin Nikolais. What seems to attract your attention? (b) Did you notice different aspects of the dance during a second viewing in comparison with your first viewing?

Seeing Color: (a) Watch the same YouTube video again. (b) As you view the video, take note of the colors. (c) What are the effects of the colors?

Noticing Contours: (a) View the same video for a third time, but this time focus on the outlines or contours of the dancers' bodies or costume shapes. (b) Did this change your experience of the dance?

Describing Angles and Lines: (a) View the same YouTube video a final time, but this time focus on the angles and lines created by the dancers' bodies or costumes. (b) How did a dancer need to position his or her body in order to create a particular angle? (c) How was a dancer positioned to create a horizontal or vertical body line?

PERCEPTION

Perception is holistic observation. Observation is the pure sensory input. Perception is a process in which neural signals and patterns are analyzed, evaluated, and transformed.[9] Incoming input is actively used to suggest ideas or test hypotheses.[10] An individual's mental expectations filter perceptions. The significance of perceptions is determined by memories, experiences, and biases.[11]

Perception requires thinking with awareness about observations so that one is conscious of that which is viewed.[12] It is possible to move through an environment and not notice or sense objects. Conscious perception develops when the environment is actively viewed as an experience. This explains why later recall of specific environmental details does not always occur.

Brain Network

Perceptions produce a network of activity in the brain that result in actions. If there is light coming from a window, a baby will be attracted to the light and turn his or her head toward the window. Perceptions precede actions.[13]

The term *action observation network* (AON) refers to a network that is more general and extends beyond the mirror neuron system in the inferior parietal and premotor cortices. The AON is composed of the supplementary motor area, ventral premotor cortex (in front of the primary motor are(a), inferior parietal lobe, and posterior superior temporal sulcus/middle gyrus[14] (see figure 2.2).

In a study of expert dancers, it was found actual physical experience is needed for a robust activation of the AON. Another study using novice dancers compared their brain activity when viewing unknown movements to viewing movements learned previously. In particular, the right precentral gyrus was activated when subjects viewed a video of a dance they physically practiced, but not activated when viewing a dance they never physically

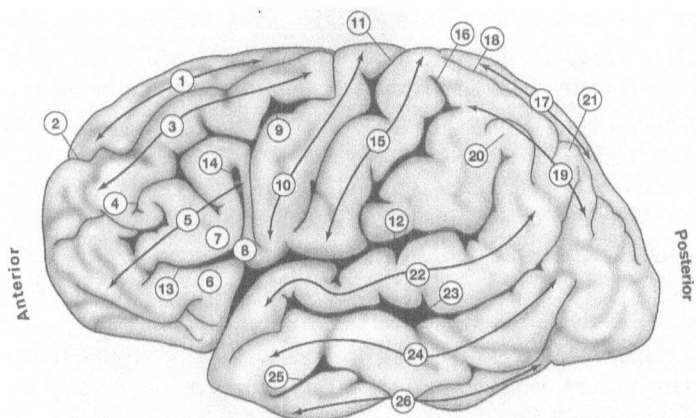

Figure 2.2   1, superior frontal gyrus; 3, middle frontal gyrus; 10, precentral gyrus; 15, postcentral gyrus; 22, superior temporal gyrus; 23, superior temporal sulcus; 24, middle temporal gyrus; 25, inferior temporal sulcus; 26, inferior temporal gyrus.
Source: Schenkman, Margaret; Bowman, James; Gisbert, Robyn; Butler, Russell, *Clinical Neuroscience for Rehabilitation*, 1st © 2013. Reproduced by permission of Pearson Education, Inc., New York, New York.

performed.[15] In addition, a larger brain area of the novice dancers was activated when they viewed an expert human model in comparison to viewing patterns of movement indicated by arrows.[16]

Classroom Applications

Some educational experts advocate using multiple senses to reach all learners because no two students learn in the same way.[17] When reading, the visual sense is engaged by calling attention to illustrations throughout a book or by putting photocopied pictures from a story into the proper order, the tactile/kinesthetic sense if students act out parts of a story, and the auditory sense if students hear a text read aloud. The Montessori system provides sandpaper letters that students trace. In science class, students can use all the senses to observe living entities and predict the outcome of experiments.[18]

Movement and Dance Connections

When students learn movements in dance class, there is variance in how individuals perceive. One student may look at spatial patterns, while another views use of the arms. Perceptions are filtered through personal biases and past experiences. The same is true of audience members. People notice different aspects of a performance.

Dancers learn movements using both passive and active observation. Copying movements without giving them much thought is passive learning. Active learners analyze details, structure, and find meaning in the movements demonstrated. They observe movement impulses, the inspiration for a dance, and analyze work from multiple perspectives.

Dance activates most of the senses. Instrumentation or vocalization can either complement or intensify movement. Some dance forms, such as percussive dance, create the rhythms and sounds through the dancers' movements. Dance is a visual and kinesthetic art form, for the audience as well as performer. A dancer's body forms many types of lines and geometric or irregular shapes that create designs between dancers or by pathways traced throughout the performance space that are perceived viscerally as well as visually (see figure 2.3).

Dancers often talk about movement textures (e.g., smooth, jagged, floaty, or described by a host of other sensory adjectives). A dancer's kinesthetic sense and accompanying proprioceptive feedback provides awareness of movement textures. Audience members have a personal empathetic, bodily response, which produces an empathetic understanding[19] and excitement, and is responsible for the enjoyment of the movements.

Figure 2.3 Geometric body shapes. *Source*: Reprinted by permission, from S.C. Minton, 2007, *Choreography: A Basic Approach Using Improvisation*, 3rd ed. (Champaign, IL: Human Kinetics). Photographer: Joe Clithero of B & J Creative Photography.

Movement Explorations

Observation and Learning: (a) Select an area of a room in your home and observe it carefully. (b) Without looking at the room, make a list of as many details as you can remember. (c) Observe the area again without the list, but this time draw relationships between body shapes or movements to what you are observing. (d) Turn away again, and make another list of observed details without looking at the first list. (e) After completing the second list, compare the two. (f) Did you remember the same details in each observation? Did the bodily or movement connection change your observations in any way?

Interpreting Sensations: (a) Select a familiar object. (b) Close your eyes and run one hand over the surface of this object. (c) How would you describe the textures you feel? (d) How might you move in response to these textures? For instance, if the object feels smooth, you might move in a continuous and fluid manner.

Passive Observation: (a) Choose a dance video. (b) Observe it passively without moving with it physically. (c) Perform movements you remember from the video.

Active Observation: (a) View the video from the above exploration, but physically perform the movements as you observe them. (b) Later, perform movements you remember from the video. (c) What is the difference in recall?

Observing through a Personal Filter: (a) After doing the above explorations, work with a partner and discuss what aspects of the dance were most prominent for each of you. (b) Did you both notice the same aspects of the dance, or did you perceive the dance differently?

RECOGNIZING PATTERNS AND RELATIONSHIPS

The human brain organizes experiences into patterns and relationships as a general principle of perception.[20] Once a pattern is recognized, there is a tendency to try to find the same pattern in new sensory information.[21] Patterns are established through repetition and from similarities, differences, or interconnections between phenomena. When looking at clouds, one notes repetitive shapes, or finds established relationships that create images such as animals.

Brain Network

When a pattern is recognized, the number of neural networks in the brain is increased and the neurotransmitter dopamine is released. It is one of the

Figure 2.4 Three round shapes with wedges missing. *Source:* From *The Tell-Tale Brain: A Neuroscientist's Quest for What Makes Us Human* by V. S. Ramachandran. Copyright © 2011 by V. S. Ramachandran. Used by permission of W. W. Norton & Company, Inc.

neurotransmitters that carries information across synapses between neurons. Dopamine is stored in the nucleus acccumbens located near the prefrontal cortex.[22] When released, it produces feelings of pleasure and satisfaction.

The visual system identifies patterns in sensory information even when parts are missing.[23] According to experts, the brain dislikes when contradictory phenomena occur simultaneously. Figure 2.4 could be interpreted as three round shapes with a wedge missing from each one. However, since the wedges are aligned as shown in this figure, a large, white triangle can alternately be perceived.[24]

Classroom Applications

Recognizing observed patterns and relationships is critical to the learning process because it is easier to remember patterns than isolated bits of knowledge. Patterns reveal connections between information that previously appeared to be unrelated.[25] Deep meaning is digested from associations between phenomena that establish and build relationships.

Patterns in sounds are experienced as rhythms. The patterns found in speech are described as an "ordered recurrent alternation of strong and weak elements in the flow of sound and silence in speech."[26] The key here is the word *recurrent*. A sound pattern is recognized through its repetition of weak and strong groupings and other key elements. Some theorists believe that both language and music create meaning for children through their structural patterns.[27]

Traditional instructional strategies that use sequential methods to introduce information do not take advantage of the brain's natural inclination to recognize patterns.[28] Students' knowledge is increased through pattern recognition that matches new information to memories, creating more extensive neural units in the brain.[29]

Movement and Dance Connections

Dance movements create patterns that build a recognizable aesthetic. Learning movement is facilitated by quickly identifying how movements are connected to form a pattern. For example, if a series of four actions are performed alternately at high and low levels, the pattern is one of contrasting levels. Tap dance, in particular, includes steps that are based on auditory rhythmic patterns. Dance teachers of many genres frequently chant the rhythmic pattern found in a movement sequence to help students remember it.

Choreographers use choreographic devices in which movement patterns are repeated, producing an array of patterns that establish relationships based on spatial designs, between body parts of individual dancers or movements of groups, variations in dynamics, and orientation to the environment. Formal choreographic devices (e.g., ABA, rondo, canon, and theme and variations) are constructed from patterns. The rhythmic intensity and aesthetic structure of a work are formed from patterns.[30]

Movement Explorations

Patterned Learning: (a) Consider the following movement sequence—four running steps, four walking steps, four running steps, two skips, four running steps, two turns, four running steps, and one jump. (b) Repeat these movements to form a pattern. (c) Does the repetition in the pattern help you remember it?

Manipulating a Pattern: (a) Use the previous movement sequence. (b) Change the facing in which you perform the pattern. (c) Change the levels in which it is performed. (d) Change its pathways and directions. (e) Connect each of your variations, and then vary the timing where appropriate. (f) How did your development of patterns affect the choreography?

Auditory Patterns: (a) Find a soundtrack of music with an established regular beat. Some examples are rock 'n' roll, Native American drumming,

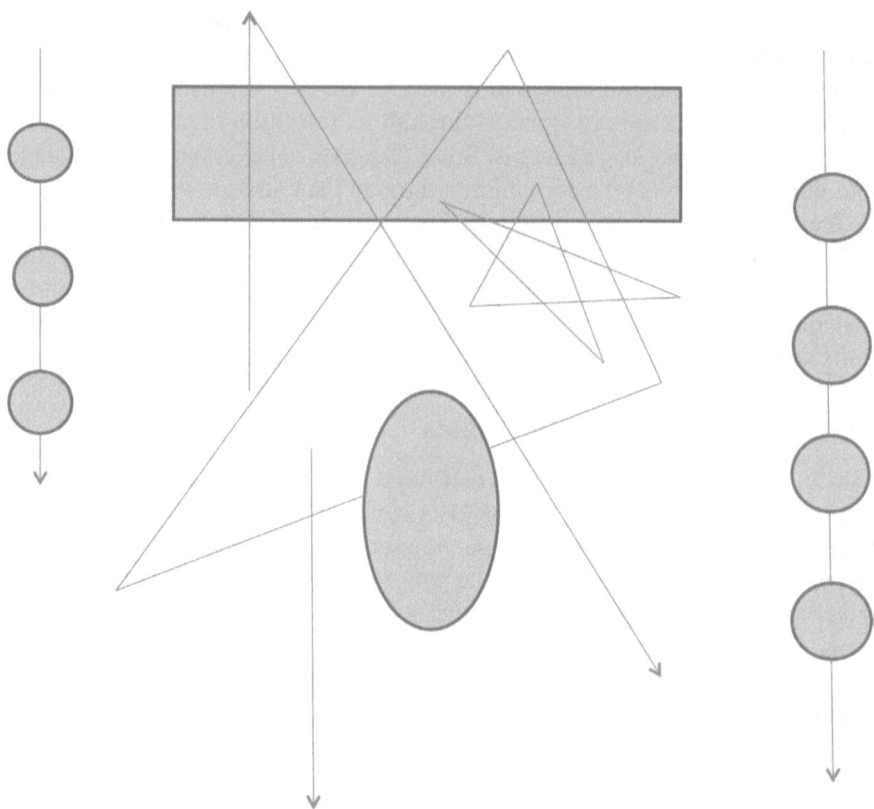

Figure 2.5 A patterned design. *Source:* Design created by S. Minton.

or marching music. (b) Create movements that reflect this regular rhythmic pattern. (c) Find a soundtrack of a polyrhythmic score. An example is African drumming. (d) Create movements based on these rhythms. (e) How did these auditory rhythms affect your movement?

Visual Patterns and Relationships: (a) Look at figure 2.5. (b) Create movement patterns suggested by some of the patterns or designs. (c) Find a relationship between these patterns and create it in movement.

NONVERBAL COMMUNICATION

Nonverbal communication transmits ideas or information without words. Approximately 93 percent of our daily emotional communication has been determined to be nonverbal, which means words contribute less than 10 percent to content communicated.[31]

Facial expressions connected to different emotions communicated through muscles of the eyes, mouth, cheeks, and neck are universal regardless of race, culture, nationality, sex, religion, or age.[32] The verbal aspects of speech consist of pronunciation style and tone. Style refers to patterns of pausing and other irregularities used by a speaker; tone is based on the pitch and volume of words.[33]

Interpersonal space or space between people talking to each other is another form of nonverbal communication. Interpersonal space is related to intimacy and varies with cultures; people from Mediterranean, Middle Eastern, or Latin cultures use closer distances between persons than individuals from the United States.[34]

In one study, researchers attempted to connect different emotions and body movements by asking subjects to view video clips of actors who portrayed angry, happy, sad, and neutral situations. Subjects were able to determine the emotion being portrayed at a level beyond chance based on use of movement quality and timing.[35]

Brain Network

Mirror neurons have been located in the ventral and dorsal premotor cortex, as well as in several regions in the parietal cortex (see figure 2.6). These brain areas are part of an action observation network (AON) mentioned earlier. Recent research suggests the mirror neuron system is particularly sensitive to observed actions and is able to produce internal motor representations that correspond to the action being observed.[36]

Mirror neurons reflect observed movements, activate circuits in the brain associated with other sensory modalities and facilitate nonverbal communication. It is possible the mirror neuron system evolved to intuit what is happening in the mind of others as an adaptive learning or social ability.[37]

Classroom Applications

Researchers explored the impact on student learning of content from PowerPoint slides when those slides were coupled with metaphoric gestures performed by the professor. The professor performed gestures matched to the meaning of the content on each slide. This study demonstrated a positive effect for the gesture-based multimedia presentation on student retention of content.[38]

Movement and Dance Connections

Movement is communicated nonverbally through observation of body use. A stooped stance communicates sadness, fatigue, or a host of other negative

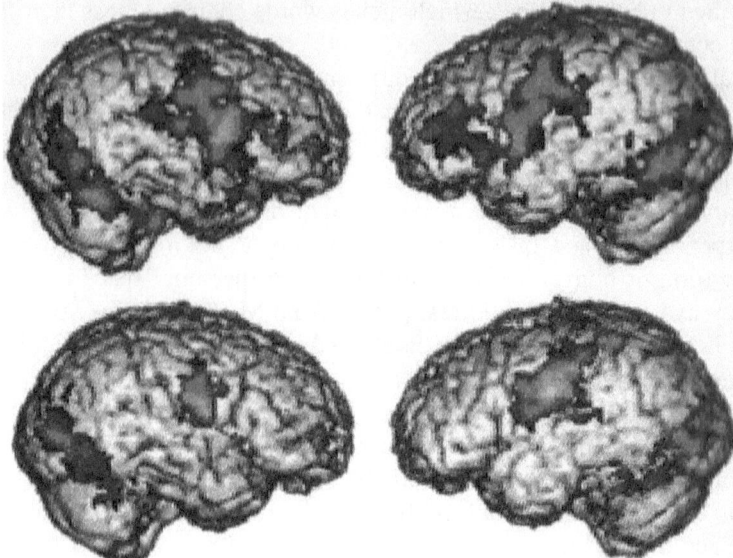

Figure 2.6 Mirror neurons noted in the dark areas. *Source: Mapping the Mind*, by Rita Carter. © 2010 by Rita Carter. Published by the University of California Press.

emotions. Moving openly with larger actions demonstrates happiness, confidence, and other positive feelings. Martha Graham is credited with her well-known statement, "The body does not lie."

Graham's solo *Lamentation* is a striking example of nonverbal communication. *Lamentation* is a portrait of a woman grieving as she sits alone on a bench. She is engulfed in material that stretches to reveal her movements as she contracts and twists her torso. "She *is* grief from the first stricken bewildered groupings of her head and torso to the last moment when she averts her covered head with a finality that is pitiful and terrible."[39]

Movement Explorations

Interpreting Nonverbal Communication: (a) Find a video on the Internet in which two people are having a conversation. (b) Mute the sound so that you can focus on the postures and gestures of the two individuals. (c) What do these postures and gestures tell you about the intent of this conversation? (d) Play the same video again, but listen to the conversation this time. (e) Did the feelings expressed by the two individuals match your interpretation of their postures and gestures when you could not hear their words?

Creating Nonverbal Communication: (a) Find photographs of people in a variety of situations and examine their body language. (b) Identify the feeling or feelings communicated by each photo. (c) Choose three photos that you think depict different messages and recreate each with your own body. (d) Does your body feel like what you thought was communicated in each photo?

Reading Nonverbal Communication: (a) Sit in an environment in which you can observe many people as they walk by. A coffee shop or airport is an ideal location for this observation. (b) Once you have found an appropriate environment, observe people as they pass by. Be especially aware of the posture, movement speed, or gestures used by the people. (c) Interpret how you believe the people you observed may be feeling or what intent they may have in mind.

Interpersonal Space: (a) Find a YouTube video of the *Moor's Pavane* created by José Limón and a second video of *Points in Space* by Merce Cunningham. (b) Watch each video carefully for any differences in the use of interpersonal space between the dancers. (c) What differences did you notice in the use of interpersonal space in these two videos? What does this communicate to you?

NOTES

1. Rita Carter, *Mapping the Mind* (Berkeley, CA: University of California Press, 2010).
2. V. S. Ramachandran, *The Tell-Tale Brain: A Neuroscientist's Quest for What Makes Us Human* (New York: W. W. Norton, 2011).
3. Carter, *Mapping the Mind*.
4. "Sight—Turning Light into Sight," Sight—Introduction—University of Minnesota, accessed November 11, 2014, http://teaching.pharmacy.umn.edu/courses/eyeAP/Eye_Anatomy/Sight/TurningLightIntoSight.htm.
5. Eric R. Kandel, *In Search of Memory: The Emergence of a New Science of Mind* (New York: W. W. Norton, 2006).
6. Emily S. Cross, "Building a Dance in the Human Brain," in *The Neurocognition of Dance: Mind, Movement and Motor Skills*, eds. Bettina Blasing, Martin Puttke, and Thomas Schack (New York: Psychology Press, 2012), 177–202.
7. Cross, *The Neurocognition of Dance: Mind, Movement and Motor Skills*.
8. Ramachandran, *The Tell-Tale Brain*.
9. Kandel, *In Search of Memory*.
10. Carter, *Mapping the Mind*.
11. Robert and Michèle Root-Bernstein, *Sparks of Genius: The 13 Thinking Tools of the World's Most Creative People* (Boston: Houghton, Mifflin, 1999).
12. Root-Bernsteins, *Sparks of Genius*.

13. James E. Zull, *From Brain to Mind: Using Neuroscience to Guide Change in Education* (Sterling, VA: Stylus, 2011).

14. Cross, *The Neurocognition of Dance: Mind, Movement and Motor Skills*; Scott Grafton and Emily Cross, "Dance and the Brain," in *Learning, Arts, and the Brain: The Dana Consortium Report on Arts and Cognition*, eds. Carolyn Asbury and Barbara Rich (New York/Washington, DC: Dana Press, 2008), 61–69.

15. Cross, *The Neurocognition of Dance: Mind, Movement and Motor Skills*.

16. Cross, *The Neurocognition of Dance: Mind, Movement and Motor Skills*.

17. Edgar McIntosh and Marilu Peck, *Multisensory Strategies: Lessons and Classroom Management Techniques to Reach and Teach All Learners* (New York: Scholastic, Inc., 2005).

18. McIntosh and Peck, *Multisensory Strategies*.

19. John Martin, *The Modern Dance* (Brooklyn: Dance Horizons, Inc., 1965).

20. Root-Bernsteins, *Sparks of Genius*.

21. Root-Bernsteins, *Sparks of Genius*.

22. Judy Willis, "The Current Impact of Neuroscience on Teaching and Learning," in *Mind, Brain, & Education: Neuroscience Implications for the Classroom*, ed. David A. Sousa (Bloomington, IN: Solution Tree Press, 2010).

23. Kandel, *In Search of Memory*.

24. Ramachandran, *The Tell-Tale Brain*.

25. Root-Bernsteins, *Sparks of Genius*.

26. Katherine Teck, *Ear Training for the Body: A Dancer's Guide to Music* (Pennington, NJ: Princeton Book Company, 1994), 123.

27. "Piaget-1 Study Music Behavior," accessed August 17, 2014, http://www.istudymusicbehavior.com/literature-review/piaget.

28. Mariale M. Hardiman, "The Creative Artistic Brain," in *Mind, Brain, & Education: Neuroscience Implications for the Classroom*, ed. David A. Sousa (Bloomington, IN: Solution Tree Press, 2010).

29. Willis, "The Current Impact of Neuroscience on Teaching and Learning."

30. Sandra C. Minton, *Choreography: A Basic Approach Using Improvisation* (Champaign, IL: Human Kinetics, 2007).

31. Albert Mehrabian, "Nonverbal Betrayal of Feeling," *Journal of Experimental Research in Personality* 5 (1971): 64–73.

32. David Matsumoto and Hyi Sung Hwang, "Facial Expressions," in *Nonverbal Communication: Science and Applications*, eds. David Matsumoto, Mark G. Frank, and Hyi Sung Hwang (Los Angeles: Sage, 2013), 15–52.

33. Mark G. Frank, Andreas Maroulis, and Darrin J. Griffin, "The Voice," in *Nonverbal Communication: Science and Applications*, eds. David Matsumoto, Mark G. Frank, and Hyi Sung Hwang (Los Angeles: Sage, 2013): 53–74.

34. David Matsumoto and Hyi Sung Hwang, "Body and Gestures," in *Nonverbal Communication: Science and Applications*, eds. David Matsumoto, Mark G. Frank, and Hyi Sung Hwang (Los Angeles: Sage, 2013): 75–96.

35. Matsumoto and Hwang, "Body and Gestures," in *Nonverbal Communication*.

36. Beatriz Calvo-Merino, "Neural Mechanisms for Seeing Dance," in *The Neurocognition of Dance: Mind, Movement and Motor Skills*, eds. Bettina Blasing, Martin Puttke, and Thomas Schack (New York: Psychology Press, 2012), 154–76.

37. Ramachandran, *The Tell-Tale Brain*.

38. Chun-Yen Chang, Yu-Ta Chien, Cheng-Yu Chiang, Ming-Chao Lin, and Hsin-Chih Lai, "Embodying Gesture-Based Multimedia to Improve Learning," *British Journal of Educational Technology* 44, no. 1 (2013): 5–9.

39. "*Lamentation* (Ballet Choreographed by Martha Graham)," accessed August 19, 2014, http://lcweb2.loc.gov/diglib/ihas/loc.natlib.ihas.200182679/default.html.

Chapter 3

Engagement

The brain is instinctively attracted to changes as a survival and adaptive necessity. Novelty is an agent for attention. New information, new experiences, and changes in the environment develop new brain networks.[1]

Engaging the mind is the crux of learning. American culture bombards students with an onslaught of stimuli. Teachers have competition from entertainment and media to capture and maintain attention and are faced with a great many questions about how to engage students so they learn effectively.

BRAIN OXYGENATION

There has been considerable research demonstrating that movement increases blood flow to the brain and stimulates brain activity. When students sit for long periods of time, blood pools in the buttocks and legs, and less circulates to the brain. Increasing the heart rate through exercise delivers more oxygen to the brain, creating conditions for more effective learning.[2]

Brain Network

In aerobic exercise, oxygen from the air is transferred to the blood through the lungs. A study done at the University of Illinois using third- and fifth-graders produced a correlation between improved academic performance and aerobic fitness.[3] The more neuroscientists discover about this process, the clearer it becomes that exercise provides an increased supply of oxygen and glucose that are an unparalleled stimulus from which the brain functions to learn.[4] This does not mean physical activity will definitely increase learning as there are many variables that determine student ability.

Any exercise will oxygenate the brain, but dance has an advantage in that it uses many areas of the brain simultaneously. Higher-level thinking skills connect the centers used in physical activity with the frontal lobe, while Broca's and Wernicke's areas in both hemispheres are required for artistic dance thought processes.

Classroom Applications

Research in physical education has supported the connection between exercise and brain function. In a study of 285 sixth-to-eighth-grade American students, it was discovered that ten minutes of rhythmic exercise in which students step on and off a low bench for three minutes was sufficient to produce improved reading comprehension in comparison to a group of similar students who did not engage in the exercise.[5]

Researchers in another American study used ninety college students to learn whether vigorous or moderate physical activity would improve students' vocabulary recall and comprehension. Engaging in vigorous exercise before or after rehearsing for a vocabulary test improved both, preparing the mind for learning and the reactive effect of learning and retrieving information.[6]

Movement and Dance Connections

The goal in an aerobic dance class is different from a normal dance class because emphasis is on developing cardiorespiratory endurance. For optimal results, such classes are structured to meet the requirements of frequency (exercise at least three times a week with twenty-four-to-forty-eight hours of rest between workouts); intensity (achieve the target heart rate required to provide best training benefits); and duration (exercise continuously at the appropriate intensity, usually for about twenty minutes).[7] In a typical dance class, participants do not perform vigorous movement continuously for twenty minutes, but any exercise will bring more oxygen to the brain, although to a lesser degree.

Many of the reasons people cite for wanting to dance reflect their delight in its physicality and health-producing effects. When the mother of a television dance personality was visiting, she said she felt tired, old, and wanted to take a nap. He suggested they try some dancing instead. Afterward, when the nap was suggested again, his mother felt energized and wanted to go for a walk or jog instead of napping.[8]

Movement Explorations

Body Energy: This exploration should be performed after you have been stationary or sedentary for a prolonged time period. (a) Note how you feel after being stationary for a long time. (b) After being stationary, get up and dance, engaging in some fairly rapid but not exhausting movement for three minutes. (e.g., you could rock out or bounce to music.) (c) Once you have completed your bout of exercise, do an assessment of how your body feels. Does your body feel different than it did before you danced? If so, in what way has your sense of your body changed?

Mental Energy: This exploration is a continuation of the one above. (d) Examine how you feel mentally. Do you feel much the same as you did before exercising, or does your mind feel different? (e) Draw some distinct comparisons between your two mental states before and after exercise.

ATTENTION

The nineteenth century philosopher William James claimed attention takes possession of the mind by choosing one of several trains of thought.[9] Cognitive scientists consider attention a sequence of mental processes allowing a person to select and concentrate on one aspect of input, while simultaneously ignoring other aspects.[10] Neuroscientists believe attention lies along a continuum of consciousness and is an executive brain function that controls behavior. From the viewpoint of embodied cognition, attention is a way to express part of the world. Sports psychologists discuss attentional style as narrow, broad, or focused.[11]

The development of focused concentration is an executive function necessary for effective learning. Executive functions of higher brain centers enable humans to reason, exercise choice, exert self-control, create, and adjust in relation to new information. Executive functions are imperative to concentration and thinking.[12]

The Brain Network

The executive function of attention depends on a neural circuit in which the prefrontal cortex is dominant.[13] The executive attention network allows humans to choose between conflicting signals by inhibiting input that might interfere with appropriate behaviors or responses. The rear section of the anterior cingulate is part of this network. It has connections to frontal and parietal brain areas.[14] In addition, the superior colliculus helps maintain activity while the brain is waiting, at rest, or in an interlude. Maintaining attention requires the brain's ability to select and increase the strength of incoming signals, while switching attention between input.[15]

All sensory input passes through the reticular activating system (RAS), a primitive network of cells located in the lower brain stem (see figure 3.1).

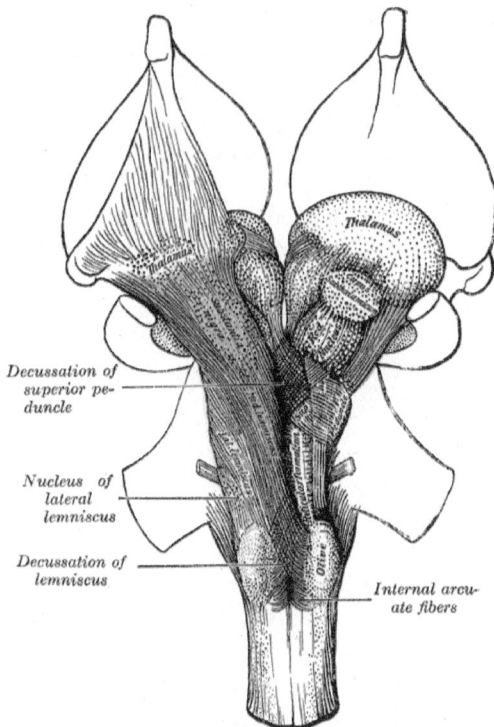

Figure 3.1 Ventral view of the recticular activating system in brain stem in center of image. *Source*: Public domain. http://en.wikipedia.org/wiki/Reticular_activating_system.

The RAS is most sensitive to critical survival information, so when a threat is perceived, it directs information to the lower reactive brain, producing an involuntary response of fight, flight, or freeze. The RAS also has a role in responding to novelty and change, and can be associated with pleasant sensory experiences that create interest and curiosity.[16]

One difficulty with maintaining attention is the brain has a limited capacity for remaining focused. While your brain has approximately one hundred billion nerve cells, only a small portion of them can be active at any moment. Overlapping patterns of neural activity compete for limited attentional resources.[17] Memory will be discussed in detail in a later chapter, but researchers have learned the inability to focus is often due to limited working memory, causing only a small amount of information to be kept in the mind while solving a problem. Some children make computational mistakes due to a lack of working memory.[18]

Some educational researchers argue that active learning is effective because it promotes enhanced attention.[19] When we are alert and paying attention, the brain's **RAS** swings into action by closing down unnecessary brain activity;

at the same time, activity is maintained in the superior colliculus, the lateral pulvinar (part of the thalamus), and the parietal cortex so that an individual can focus.[20] With the appearance of appropriate lesson content, an alert brain springs into action to facilitate learning.[21]

Problem solving intensifies attention, although students who answer questions risk making a mistake. Risk taking is essential in learning. Mistakes cause levels of dopamine, a neurotransmitter, to fluctuate in the brain, simultaneously increasing the brain's receptivity to learning a correct answer later. When corrective feedback is offered immediately, the brain attempts to alter incorrect information in a targeted neural network so the same mistake is not made again.[22] Dopamine and serotonin are neurotransmitters that produce positive feeling, and the genes responsible for their transmission influence executive attention.

Classroom Applications

As mentioned earlier, the brain is attracted to novelty. Suggested novelty events include changing the arrangement of a classroom, providing new wall displays, including different colors in lesson materials, advertising upcoming units of study, and using dance, costumes, music, and other techniques that stimulate student curiosity.[23]

Early in the past century, John Dewey proposed active pedagogical practices. This led to the acceptance that content is remembered more quickly and retained readily from learning through experience as opposed to rote and lecture-based teaching formats. Rote learning produces mental storage of materials that has little connection to existing cognitive structures in the mind. Experiential learning and problem solving enable students to make personal connections with prior knowledge, and change their thinking through reflection. It inspires creative application of information.[24]

Children are naturally curious, and when they are allowed to follow their interests, pleasure in learning increases. During a study with third-grade Italian children, the students built a model of a classroom they were to move into. Through students' verbalizations and observation of their nonverbal behaviors, the study revealed student curiosity is complex and multifaceted.[25] The researchers recommended putting students in situations with real study materials so it can be determined how they problem solve to understand content.[26]

Each student has a unique network of neurons in their brain based on genetics and experience. Thus, learning is a highly personal process. The variety in brain networks and personal experiences that exist among students in a single classroom presents an ongoing challenge for teachers. Active, experiential learning provides opportunities for students to connect prior and new knowledge in a way that suits individual needs of diverse learners.

Until recently, it was believed learning was a linear process, somewhat like moving up a ladder from one step to the next. More recently, cognitive scientists have described learning as a web with multiple pathways or independent strands. Individual skills split off and merge as needed in each child's brain, even though students may pass through the same developmental levels.[27]

Learning is not a simple endeavor, but more like work on a construction site in which the information learned is built up, transformed, and tested as students develop unique understandings.[28] A South Korean study of two groups of sixth-graders showed a constructivist teaching approach was more effective than traditional teaching methods. In this study, constructivism was described as an active process in which knowledge is formed from one's sensory and perceptual experiences to create personal understanding of content in a new cognitive framework.[29]

Traditional factory-model curricula are based on assumptions that knowing facts creates understanding. Conceptual knowledge is a higher cognitive level. It connects simple and more complex brain processing centers, allowing the mind to operate simultaneously on several cognitive levels.[30] A factual question is: What is the climate of the Amazon? This question would become conceptual by asking: What are some other locations in the world that have the same type of climate? This change transfers information from one situation to another through comparison, evaluation, and selection.

Students enter the classroom with many social, familial, and economic pressures that produce problems that interfere with focus. Teachers are facing issues of students' attention in critical or catastrophic proportions. A strong student-teacher interrelationship and high relevance of class content to a student's reality help modify distracting factors. If students believe a teacher is concerned, impartial, respectful, and presents content relevant to their identity, interests, and future goals, they are encouraged to pay attention.[31]

Group work and interpersonal communication have been suggested as ways to improve student attention. In group work, researchers found it was important for group members to listen and respond to ideas expressed by others. This can be accomplished by having one group member echo ideas presented by others or ask clarifying, probing questions.[32] Clarifying questions make students think about reasons for their statements.[33]

Movement and Dance Connections

A college dance education class observed elementary school children during recess. The observations demonstrated that children are creative as they play, using similar movements as those used in creative dance. Another group of dance education students observed elementary students who were asked to sit quietly at their desks and found that children are naturally physically expressive and have problems sitting quietly for prolonged periods. These observations endorsed the idea that children who are allowed to move are creative and more focused.[34]

Creative movement provides opportunities for change and novelty. The high-action level excites expectation and promotes engagement through movement variety. When taught creatively, learning dance heightens critical thought and problem-solving experiences.

Learning movements in dance skills or technique class is also a form of active, experiential learning, and as such it promotes focus and attention. It is impossible to learn a movement sequence without being aware of how the sequence begins, develops, and ends. Learning movement also means becoming aware of subtleties and changes as they occur.

As mentioned previously, learning involves connecting new content to what is known. Students in dance class enter with a variety of past experiences. Learning is facilitated when new materials connect to recognizable movements.[35] Dance teachers need to emphasize these connections. For example, dance students can learn to perform a turn in the air by connecting this action to an understanding of how to turn on the ground.

Dance making is often motivated by curiosity. Curiosity can capture a choreographer's interest, stimulate observation, and spur investigation in subject matter that inspires rich, varied movements. It is necessary to recognize and follow creative impulses to discover movement possibilities.

Dances that present a novel approach are usually successful and capture audience interest. Audience members respond to the visual appearance of movements, placement in space, energy quality, timing, and kinesthetic resonance. Watching a dance performance is also an experience in making connections with one's past, present, and future in a way that is special for each person. Experiences leave a neurological record that changes the brain. When viewing a dance, audience members relate to it in reference to their individual neuro network. This inspires interest while providing freedom of interpretation.

Movement Explorations

Novelty: (a) Watch a dance video and note movements that were novel to you. (b) Explain why the movements were novel. (c) Create a movement sequence that is novel to you.

Replication: (a) Locate a YouTube video of a short Flash Mob dance sequence. (b) Observe and replicate a movement sequence with awareness of what you are noticing and repeating. (c) What did you notice and learn immediately, and what took you longer to learn? Analyze why.

Attention to Detail: (a) Watch the same video you used in the preceding exploration. (b) View it several times, each time focusing on a different section of the body and adding it to the performance of the movement or sequence: the feet and legs, the center of the body (torso), the arms, and head. Last, notice the visual focus. (c) What effect did use of a targeted focus have on your attention span? What effect did attention to detail have on learning this dance movement or sequence?

Constructivist Learning: (a) Find another YouTube video of a simple dance movement or sequence. (b) Decide how you will go about learning it. Would you begin with the feet, or would you begin with spatial directions? (c) Once you have created a method for learning, use this method to learn the movement demonstrated so that each subsequent step in your learning process builds on those that came before. (d) Was this method of learning successful? Did your personal construction of your learning engage your attention?

Personal Approach: (a) Have a friend learn the same dance movement or sequence you learned in the preceding exploration. (b) Discuss the techniques you each used to learn it. (c) Did the two of you use the same learning techniques, or were your strategies different?

MOTIVATION AND ENGAGEMENT

Motivation is affected by the brain's reward system, a sense of freedom, mental imagery, involvement in creative work, and having an uncluttered mind. Researchers have been able to link the arts with motivation and engagement. The first step in this link is interest.

Daniel Pink, noted champion for arts in education, proposes three major elements to maximize motivation that run contrary to common business and educational practices: autonomy, mastery, and purpose.[36]

Autonomy permits workers or students to determine what they wish to achieve on their own schedule, with their own methods of achievement. When companies release workers from supervision, employees are more engaged, energetic, and creative. Spurred by self-created challenges, people are stimulated to pursue discoveries they would not pursue in a directive environment. Data has driven the conclusion that humans have an innate drive to be autonomous, self-determined, and connected. When that drive is liberated, people achieve more.[37]

Mastery is a combination of "flow,"[38] and the elation when things are working smoothly, energy is high, and ideas are streaming. The effort, energy, work, and practice required to achieve ideas is surmounted. Inner goals rather than extrinsic assignments are the driving force. Achieving excellence requires perseverance, but the will to work toward accomplishment brings self-fulfillment when it is an inner goal.

A goal without a purpose will not motivate workers or learners. Substantial examples of data from systems of rewards and punishments versus providing people with a higher purpose demonstrate that employees and students work harder and longer when they understand and believe in the greater meaning of their efforts. Goals are often specific accomplishments, but "purpose" encompasses a mission on a higher level. Scientific findings about motivation seem to directly contradict common practice.[39] Students in fine arts programs report that when involved in a creative project they have a commitment to the task and are motivated to finish projects.[40]

Brain Network

Motivation and attention involve a neurological relay system. The cerebellum, a relatively primitive part of the brain, governs and refines movement when we learn something physical; it also ensures that brain systems run smoothly and updates and manages the flow of information. In turn, the cerebellum passes information to the basal ganglia and then to the prefrontal and motor cortices; the basal ganglia act as a sort of automatic transmission system that can shift resources for attention in line with cortical demands, while the prefrontal and motor cortices are the centers for thinking and movement, respectively.[41]

Motivation is a function of brain signals traveling across intact nerve pathways. Remaining engaged is dependent on how fluidly information travels across those pathways.[42] Movement is controlled by these same networks and neurologically excites engagement.

Dopamine enhances motivation and is produced through exercise and motor activity. It carries signals to the brain's reward center, which must be sufficiently activated in order to send a message to the prefrontal cortex to make the signals high priority and to attract attention. The prioritizing aspect of executive function is an essential component of motivation.[43]

A neuroscience study on creativity focused on jazz musicians who received fMRI imaging while they were involved in spontaneous improvisation. During improvisation, the lateral prefrontal cortex of the musicians was deactivated; normally the prefrontal cortex is associated with regulation, focused attention, inhibition, and monitoring oneself.[44]

Classroom Applications

Researchers postulate that children who have a high level of openness and a normal level of interest will have high motivation to receive training in an art form because their engagement is sustained over longer time periods.[45] Students are also motivated when they have a personal mental image of their goals or solution to a problem. Students become engrossed in a creative project, and their work begins to flow.[46]

There are a variety of conditions that build motivation and enhance engagement. Interest is sustained, rewarded, and even fun as long as progress continues toward a goal; freedom in an educational environment provides intrinsic motivation for students—a condition that is difficult to find in most schools.[47] Much of a teacher's time is devoted to controlling students' behaviors, although better behavior could be engendered by allowing students the freedom to work on projects of their own choice in relation to a prescribed curriculum.

Movement and Dance Connections

Motivation and engagement have physical roots. Exercise produces dopamine and feelings of satisfaction, which means movement-based teaching strategies are brain compatible. Research has shown it is effective to include the whole body in the learning process.

A Neurological Reorganization Therapy Movement Program was used to test its effect on ten students ages five to nine, seven of whom had learning disabilities. A Brain Dance warm-up was followed by crawling, creeping through an obstacle course, and cool-down. The Quick Neurological Screening Test II (QNST-II) was used pretest and posttest to measure disorders of attention, among other student capabilities. The teachers and parents recorded student behaviors in a daily log, and the researcher kept a daily journal. Posttest scores on the QNST-II decreased for nine of the ten students, indicating a reduced struggle with daily activities.[48]

Making a dance is a creative quest. Mature work is fluent because the choreographer approaches an inspiration from a number of viewpoints and perspectives in a way that is unique. Commitment to the inspiration for a work drives a choreographer to see a dance through to completion. As more of a dance work is revealed, it is as if the choreographer is riding a wave and caught in the momentum of each moment.

The "high" of dancing hooks students. While performing, a dancer frequently experiences being fully present.[49] Life feels intensified. This achievement plus the prospect of audience recognition can motivate a person to sustain difficult training. A student asked Martha Graham, "Should I become a dancer?" Graham responded, "If you have to ask, the answer is no." Becoming a dancer is not a logical choice, but a passion. Learning dance technique is a long, hard road, requiring fortitude, persistence, devotion, and self-motivation. Unification of mental plus physical involvement creates a "flow" that drives dancers almost past endurance to achieve the divine.

Movement Explorations

Freedom of Choice: (a) Do whatever you want in movement for twenty minutes. (b) Be aware of your process and the directions of your thinking. What were the choices you made? (c) Note your interest level. (d) Was the outcome satisfying?

Uncluttered Mind: (a) Find a teaching YouTube video that demonstrates how to perform a simple movement or step. (b) As you watch this video, think about a number of tasks that you need to complete for the week. Then, attempt to perform the movement being demonstrated. (c) Watch the video again, but think only about the step being demonstrated and attempt to physically perform it. (d) Was there a difference in your ability to perform this step during the second practice session?

Attention and Focus: (a) Find a teaching YouTube video of another step or movement. (b) As you watch the video, focus solely on one aspect of the movements that catches your attention, such as the way the dancer moves around the space. What about it interests you? (c) Watch the video again, but this time pay attention to a different movement aspect, such as changes in timing. (d) What was your reaction to the movement after your first viewing? Was your perspective different after the second viewing? Describe the differences and similarities between the first and second viewing. (e) Did changing the focus of your attention sustain your interest?

NOTES

1. Adele Diamond, "Want to Optimize Executive Functions and Academic Outcomes? Simple, Just Nourish the Human Spirit," *Minnesota Symposia on Child Psychology: Developing Cognitive Control Processes: Mechanisms, Implications, and Interventions* 37 (2014): 205–30.

2. Traci Lengel and Mike Kuczala, *The Kinesthetic Classroom: Teaching and Learning through Movement* (Thousand Oaks, CA: Corwin, 2010).

3. John J. Ratey, *Spark: The Revolutionary New Science of Exercise and the Brain* (New York: Little Brown, 2008).

4. Robert Brooks, "Physical Exercise in School: Fitness for Both Body and Mind" (presentation, Learning and the Brain Conference, Washington, DC, May 7–9, 2009).

5. Tim Mead, Susan Roark, Lane J. Larive, Kristen Percle, and Rachel N. Auenson, "The Facilitative Effect of Acute Rhythmic Exercise on Reading Comprehension of Junior High Students," *Physical Educator* 70, no. 1 (2013): 1–16.

6. Andrea S. Salis, "Proactive and Reactive Effects of Vigorous Exercise on Learning and Vocabulary Comprehension," *Perceptual & Motor Skills: Motor Skills & Ergonomics* 116, no. 3 (2013): 918–28.

7. Esther Pryor and Minda Goodman Kraines, *Keep Moving! It's Aerobic Dance*, 3rd ed. (Mountain View, CA: Mayfield Publishing, 1996).

8. Derek Hough, *Taking the Lead: Lessons from a Life in Motion* (New York: William Morrow, 2014).

9. William James, *The Principles of Psychology*, Vol. 1 (New York: Henry Holt, 1890) 403–4.

10. John R. Anderson, *Cognitive Psychology and Its Implications*, 6th ed. (New York: Worth Publishers, 2004).

11. Glenna Batson with Margaret Wilson, *Body and Mind in Motion: Dance and Neuroscience in Conversation* (Chicago: Intellect, 2014).

12. Diamond, "Want to Optimize Executive Functions and Academic Outcomes?"

13. Diamond, "Want to Optimize Executive Functions and Academic Outcomes?"

14. Michael I. Posner, "Neuroimaging Tools and the Evolution of Educational Neuroscience," in *Mind, Brain, & Education: Neuroscience Implications for the Classroom*, ed. David A. Sousa (Bloomington, IN: Solution Tree Press, 2010), 27–43.

15. Helen Neville, Annika Andersson, Olivia Bagdade, Ted Bell et al., "Effects of Music Training on Brain and Cognitive Development in Under-Privileged 3-to-5 Year Old Children: Preliminary Results," in *Learning, Arts, and the Brain: The Dana Consortium Report on Arts and Cognition*, eds. Carolyn Asbury and Barbara Rich (New York/Washington DC: Dana Press, 2008), 105–16.

16. Judy Willis, "The Current Impact of Neuroscience," in *Mind, Brain, & Education: Neuroscience Implications for the Classroom*, ed. David A. Sousa (Bloomington, IN: Solution Tree Press, 2010), 44–66.

17. V. S. Ramachandran, *The Tell-Tale Brain: A Neuroscientist's Quest for What Makes Us Human* (New York: W. W. Norton, 2011).

18. Daniel Ansari, "The Computing Brain," in *Mind, Brain, & Education: Neuroscience Implications for the Classroom*, ed. David A. Sousa (Bloomington, IN: Solution Tree Press, 2010), 201–25.

19. James P. Byrnes, *Minds, Brains, and Learning* (New York: The Guilford Press, 2001).

20. Rita Carter, *Mapping the Mind* (Berkeley: University of California Press, 2010).

21. Carter, *Mapping the Mind*.

22. Willis, "The Current Impact of Neuroscience."

23. Willis, "The Current Impact of Neuroscience."

24. Greta G. Freeman and Pamela D. Walsh, "You Can Lead Students to the Classroom, and You Can Make Them Think: Ten Brain-Based Strategies for College Teaching and Learning Success," *Journal on Excellence in College Teaching* 24, no. 3 (2013): 99–120.

25. Maria V. Cifone, "Questioning and Learning: How Do We Recognize Children's Questions?" *Curriculum and Teaching Dialogue* 15, nos. 1 & 2 (2013): 41–55.

26. Cifone, "Questioning and Learning."

27. Kurt W. Fischer and Katie Heikkinen, "The Future of Educational Neuroscience," in *Mind, Brain, & Education: Neuroscience Implications for the Classroom*, ed. David A. Sousa (Bloomington, IN: Solution Tree Press, 2010), 249–69.

28. Fischer and Heikkinen, "The Future of Educational Neuroscience."

29. Jong Suk Kim, "The Effects of a Constructivist Teaching Approach on Student Academic Achievement, Self-Concept, and Learning Strategies," *Asia Pacific Education Review* 6, no. 1 (2005): 7–19.

30. H. Lynn Erickson, *Concept-Based Curriculum and Instruction for the Thinking Classroom* (Thousand Oaks, Corwin Press, 2007).

31. Michael J. Corso, Matthew J. Bundick, Russell J. Quaglia, and Dawn E. Haywood, "Where Student, Teacher, and Content Meet: Student Engagement in the Secondary School Classroom," *American Secondary Education* 41, no. 3 (2013): 50–61.

32. Ron Ritchhart, Mark Church, and Karin Morrison, *Making Thinking Visible: How to Promote Engagement, Understanding, and Independence for All Learners* (San Francisco: Jossey-Bass, 2011).

33. Ritchhart, Church, and Morrison, *Making Thinking Visible*, 37.

34. Karen Kaufmann and Becky Ellis, "Preparing Pre-Service Generalist Teachers to Use Creative Movement in K–6," *Journal of Dance Education* 7, no. 1 (2007): 7–13.

35. Alireza N. Moghaddam and Seyed M. Araghi, "Brain-Based Aspects of Cognitive Learning Approaches in Second Language Learning," *English Language Teaching* 6, no. 5 (2013): 55–61.

36. Daniel Pink, *Drive* (New York: Riverhead Books, 2009).

37. Pink, *Drive*.

38. Mihaly Csikszentmihalyi, *Flow* (New York: Harper & Row, 1990).

39. Charles J. Limb and Allen R. Braun, "Neural Substrates of Spontaneous Musical Performance: An fMRI Study of Jazz Improvisation," *Library of Science One* 3, no. 2 (2008): 1–9.

40. Eric Jensen, *Arts with the Brain in Mind* (Alexandria, VA: Association for Supervision and Curriculum Development, 2001).

41. Ratey, *Spark*.

42. Ratey, *Spark*.

43. Ratey, *Spark*.

44. Limb and Braun, "Neural Substrates of Spontaneous Musical Performance: An fMRI Study of Jazz Improvisation."

45. Michael Posner, Mary K. Rothbart, Brad E. Sheese, and Jessica Kieras, "How Arts Training Influences Cognition," in *Learning, Arts, and the Brain: The Dana Consortium Report on Arts and Cognition*, eds. Carolyn Asbury and Barbara Rich (New York/Washington, DC: Dana Press, 2008), 1–10.

46. Byrnes, *Minds, Brains, and Learning*.

47. James E. Zull, *From Brain to Mind: Using Neuroscience to Guide Change in Education* (Sterling, VA: Stylus Publishing , 2011).

48. Megan Sissom, "Neurological Reorganization Adapted for the Classroom, and Learning Disabilities" (master's thesis, Antioch University, Seattle, n.d.).

49. Thomas Schack, "Building Blocks and Architecture of Dance," in *The Neurocognition of Dance: Mind, Movement and Motor Skills*, eds. Bettina Blasing, Martin Puttke, and Thomas Schack (New York: Psychology Press, 2012), 11–39.

Chapter 4

High-Level Thinking Skills

As early as 1956, domains of cognition were defined and rated in a hierarchy of educational goals by Benjamin Bloom in his book popularly called *Bloom's Taxonomy*.[1] In the new millennium, the taxonomy was adapted to contemporary understanding by Anderson and Krathwohl. In ascending order, Bloom's taxonomy ranks thinking skills using nouns: knowledge, comprehension, application, analysis, synthesis, and evaluation. Thinking is an active state of mind, so Anderson and Krathwohl used verbs: remember, understand, apply, analyze, evaluate, and create.[2]

CATEGORIZATION, ORDER, AND ORGANIZATION

Categorization creates order and organization by recognizing similarities, comparing differences, contrasting, and sequencing. It is a form of abstract thought that requires high-level cognitive skills to distinguish similarities and differences, recognize patterns, conceptualize abstract qualities, and understand a progression.

Brain Network

Damage to Broca's area destroys the ability to categorize. In order to sort ideas or objects into categories one must abstract, interpret, or infer their general nature, a function of the left hemisphere. As the dominant locality for expressive language, Broca's area plays a major role in the process of abstraction.[3] Meaning and abstraction extract the essence of an entity or idea and are essential to categorization.

Working memory should be mentioned here due to its function in ordering and organizing objects or ideas. Working memory, discussed in chapter 6, enables information to be held, manipulated, and organized in the mind while solving problems.[4] It involves a number of brain areas, including the visual occipital lobe, auditory temporal lobe, body connections in the parietal lobe, and executive thought in the prefrontal cortex. All of these areas are therefore active in the orchestration of categorization.

Classroom Applications

The ability to sort information is central to learning. It is the basis of learning to read, performing calculations, understanding data, and making sense of one's environment. Poor organizing skills make it impossible to realize goals. Unless progressive steps toward an aim are understood and followed, the desired end result is not achieved.

Categorization facilitates recall of materials and events identified by their qualities. For example, items recalled about gardening such as daisy, peas, and shovel are easier to remember if grouped according to their respective categories—flowers, vegetables, and garden tools. Experts also indicate recall is improved if students organize content themselves.[5]

Studies demonstrated a connection between order and the mental representation of numbers. Students form a spatial image of numbers positioned on an imaginary ruler in the brain.[6] When people of a variety of ages have been asked which of two numbers is larger, they are able to answer more quickly if there is greater distance between numbers. This suggests it is easier to compare two numbers that are farther apart on an imaginary number line.[7]

Organizational skills are essential for language acquisition. In their first year, babies organize sounds into patterns, and connect the sounds to words and to objects words represent.[8] In the second year, understanding words leads to linking words together in sequences, and then grammatical sentences. It has been theorized that the human brain with its larger frontal lobe is innately programmed to learn language because it organizes grammar quickly.[9] Deaf babies with deaf parents follow a similar process by producing hand movements as a subset of their parents' sign language.[10]

It has been shown that student GPAs correlate with the ability to organize. Researchers learned that a binder used to organize lessons had a positive effect on learning for middle school students with Attention-Deficit/Hyperactivity Disorder (ADHD). This executive function difficulty affected students' ability to focus and organize their thoughts. This study demonstrated systematic binder organization was a better predictor of student improvement than more therapeutic interventions.[11]

Movement and Dance Connections

Categorization of movement defines specific genres of dance. Tribal dance relates to the earth and spiritual life. Ballet grew from the graceful affectations of court dance. Modern dance, in rebellion to the artificiality of ballet, uses personal expression, gravity, and articulation of the spine. Tap developed from a conglomerate of cultures with a focus on rhythms and sounds. Contemporary urban society is expressed in hip hop, which is fast paced, gymnastic, and competitive.

Specific movements fit certain categories. Leaping and jumping are elevated movements that leave the ground. Running, galloping, or crawling are locomotor movements that travel through space. Bending and stretching are nonlocomotor, axial movements that remain in one spot.

Laban's Effort Shape analysis system mentioned in chapter 1 organizes bodily movement into relationships with space, time, and energy. It has become an extensively taught system of movement analysis, especially in higher education.

Dance classes follow a structured organization in a progression for healthful body use. Movements with like attributes are learned during separate sections of a class. When planning classes, dance teachers are mindful of creating a lesson based on body needs and a required movement order. Usually the first part of each class is organized to give students an opportunity to warm up. This section leads to movements in complex sequences, then to movements across and around the space, with time for cooling down at the end of class.

In contrast, choreography is organized as a communicative flow since different types of movements convey contrasting feelings to an audience. Slow, continuous actions express calm. Jumps project joy or excitement. Successful dance works provide a strong overall structure and are shaped into a continuum from beginning to end. It is a choreographer's job to organize the sequence of movements and progression in a dance.

Organizational skills also come into play when directing a performance. The countdown to a performance begins by creating a schedule for rehearsals, first in available studio spaces and later in the performance venue. Promotional aspects, music, costumes, sets, props, lighting, and technical needs are budgeted and arranged. Once in the theatre, blocking, lighting, and the dress rehearsal finalize arrangements. Each step supports those that follow to fulfill an artistic vision.

Movement Explorations

Categorize: (a) Find a dance on YouTube that has musical accompaniment with a steady rhythm, and note movements that are performed at different levels—high, middle, or low. You may need to watch the dance more than once in order to remember the various actions. (b) Organize all of the movements together that were performed at a high level, all of those together performed at a middle level, and all of those together performed at a low level. (c) Were there other characteristics you could use to create new movement categories?

Organizing Space, Time, and Energy: (a) Use the same dance video, but this time focus on one dancer in the piece and how he or she uses space, time, and energy. (b) After viewing the dance several times, draw a diagram of the pathway in space followed by your selected dancer. (c) Select a movement you find interesting and count the number of beats it takes for the dancer to perform the movement. (d) List the energies (efforts) mentioned in Laban Analysis, chapter 1. (e) Develop a sequence following a pathway in your diagram, using the number of counts in your selected movement and a contrasting set of energies.

Ordering: (a) Choose movements from explorations 1 or 2. (b) Place the movements in an order that seems interesting to you and perform them. (c) Reverse the order of these same movements and perform the order backward. (d) What issues did you encounter during this process?

CRITICAL-THINKING SKILLS

Bloom's hierarchical critical thought processes outlined in the beginning of this chapter are essential elements of education. It is hoped that students learn higher-level thinking skills so they can apply understanding to life situations. Reading texts for literal or factual information without exploring deeper meaning does not develop understanding. When students merge their thinking with content they make connections, question, infer, visualize, and determine the importance of information.[12]

Brain Network

Traditionally, neural mechanisms that control movement and higher-level thinking were considered separate and discreet. Recent evidence shows motor coordination and cognition are connected.[13]

Current models of the brain extend language processing to a diverse cortical network. Broca's and Wernicke's areas are integral to language formation and comprehension, especially to verbal and semantic processing. Semantic processing also involves the temporal gyrus. This means the left temporal lobe is key in comprehending spoken language. Broca's area also connects with working memory (see chapter 6), or the ability to hold content in mind to analyze, manipulate, apply, and evaluate it.[14]

As previously mentioned, patients with damage in Broca's area can understand and read but cannot reconstruct their thoughts to express them. People with damage in Wernike's area cannot comprehend language and are unable to read. In order to verbalize thoughts or understand what is being read, the reader makes inferences based on prior knowledge found in the brain's semantic networks, along with answering questions, imaging, predicting, clarifying, and summarizing.[15]

In a positron emission tomography (PET) study, adults read passages they could not understand. When they were given a contextual clue that helped them comprehend passages, numerous brain regions were more active than when the subjects read the passages without comprehension. The added brain activity was found in the middle, lower part of the frontal cortex behind the eyes, with added activity in other parts of the brain. This part of the frontal cortex is associated with the brain's reward network, which may indicate reading with comprehension is rewarding.[16]

Classroom Applications

Critical thinking is essential to survival, but pedagogy has not always supported it. Socrates understood two thousand years ago that meaningful questions evoke complex thought. They stimulate the mind to remember, analyze, and explore solutions. Teachers can encourage deep thinking in students by asking meaningful questions.

To reason well is essential for clarity, accuracy, relevance, depth, breadth, logic, significance, and fairness. Clarity requires elaboration, good examples, and illustration. Accuracy means checking results for truthfulness. Relevance is whether thinking relates to the problem at hand. Depth refers to the complexity and difficulty. Breadth indicates an ability to consider issues from multiple perspectives. Logic involves thinking that makes sense. Significance depends on choosing the most important questions to be answered. Fairness means considering other viewpoints.[17]

Researchers studied whether or not critical-thinking skills were promoted through construction of concept maps. Concept maps were created through line drawings or pictures that depict complex, abstract content, illustrating connections between parts. It was found that the creation of concept maps facilitated critical thinking in some, but not all, cases. They were most effective when content included dynamic processes, coherence, and interrelationships between subtopics.[18]

Movement and Dance Connections

The arts are particularly effective for stimulating critical thinking because each problem has multiple solutions. Dance enables students to make new connections, refine their thinking, and make contributions to human perceptions and perspectives through the medium of movement.[19]

Dance exercises the critical-thinking skills outlined in Bloom's taxonomy.[20]

Knowledge/Remember: Imitating, replicating, or recalling movement without deep thought is the equivalent of memorizing knowledge; a form of parroting motion that is akin to rote learning. Dance is physically active, but most classes do not demonstrate the understanding acquired through active learning.

Comprehension/Understand: The reflective dancer perceives techniques and movements with understanding and purpose. Movement is a language that communicates values, beliefs, and ideas. Even dances that are purely entertainment or recreational reflect the culture from which they emerge.

Application/Apply: The performing of dance movements or audience enjoyment of a dance embodies an artistic intent. This gives the dance a practical and functional existence. Although dance as an art form is an abstraction of everyday practical movement, the practice of dance is the application of its aesthetic and meaning.

Analysis/Analyze: Probing deeper into knowledge reveals its form or structure. In dance technique class, movement is analyzed to learn how the body is used to replicate movement traits. Choreography is analyzed for compositional structure, qualities, and meaning. Analysis takes many forms, but without it the audience receives little understanding, although the experience may be enjoyable.

Synthesize/Evaluate: These terms are not equivalent with identical parallel meanings. Bloom's purpose in synthesis is to integrate ideas into a unified whole. To evaluate is to give importance or create a hierarchical valuation. Nevertheless, a completed dance is a synthesis that is frequently evaluated or critiqued.

Evaluate/Create: Bloom's highest level of critical thought is "evaluate," whereas the contemporary taxonomy places importance on creativity as its culmination. The creative act is like an explosion of new ideas built from learned knowledge. Some dance teachers believe students cannot choreograph until they have mastered technique, but a very young child dancing is akin to scribbling with the body. As greater movement skills and control are acquired, movements become purposeful and choreography is structured through higher-level thinking.

The goals of dance education are to: (1) improve movement ability; (2) view and critique dance; (3) create dances; and (4) learn about the history and culture of dance. Curriculum that fosters critical thinking is student centered, experiential, and one in which students construct understanding of dance concepts using related movement analyses.[21]

Movement Explorations

Knowledge/Remember: (a) Find a simple movement sequence on YouTube. (b) Repeat it until you have learned it. (c) The next day, without viewing the sequence, perform it from memory.

Comprehension/Understand: (a) With another person, watch part of a dance created by one of the pioneers of modern dance such as Martha Graham. (b) Without discussing the dance, share each of your interpretations of the dance's meaning. (c) Discuss the similarities and differences of interpretation and reasons supporting them.

Application/Apply: (a) Select one interpretation of the dance you viewed in the preceding exploration. (b) Create several movements that communicate the same meaning.

Analysis/Analyze: View the same dance you viewed in the preceding exercises, but this time analyze how the dance communicated its meaning through its movement and structure.

Synthesis/Evaluate: (a) Examine how the dance you viewed is unified into a work of art with meaning. (b) Evaluate how well you think the different parts of the dance fit together. Are the transitions smooth? Is the development progressive? Does the ending provide a sense of completion?

Evaluation/Create: (a) Does the dance inspire new movement ideas for you? (b) Improvise your own movement based on the ideas and movements in the dance you viewed. (c) Develop movement based on your own ideas.

FLEXIBILITY AND ADAPTABILITY

Adaptability is essential for survival. To arrive at conclusions with creative scope requires a person remain open to possibilities, consider new information, play with alternative viewpoints, and discover new input through intuition and flexibility.[22] Creative approaches encourage open minds and stimulate multiple perspectives, conclusions, or responses to a question.[23]

Brain Network

Brain plasticity is a response to the need to adapt. When part of the brain is injured or nonfunctional, new neurons connect with other areas of the brain to serve a necessary function. If an infant is blind beyond the critical visual period (three to eight months), neurons for touch might connect with visual centers to intensify textural information and provide spatial images.[24] If a stroke destroys brain areas, the brain responds by generating new neurons and by reorganizing across large brain sectors.[25]

Environmental factors perpetually change, as does human adaptation. The brain quickly recognizes change and reorganizes thoughts to accommodate. A person's response through movement to environmental changes is reactive control, while an internal system adjusts under central control. Both demonstrate their ability to adapt.[26]

Classroom Applications

Flexibility was studied in young children by supplying toddlers with balls, ramps, trays, and a variety of containers that they were allowed to freely explore in supervised play. Without adult direction, the toddlers investigated the materials by rolling, balancing, manipulating, making sounds, filling and dumping content, transporting them, and often copying the actions of another child. Through this study, the researchers gained an appreciation of the flexible thinking displayed by young, unschooled subjects.[27]

Flexible learning, in which college students used a personalized approach and were allowed to make choices about where, when, and how their learning occurred, correlated with higher achievement. Students who got the best scores on achievement tests were those who followed a flexible learning schedule.[28]

Movement and Dance Connections

In order to achieve physical skills, dance technique requires mental analysis as well as physical flexibility to adapt mind and body beyond personal physical limitations. Performing is an environmental challenge. Falls, bumps, mistakes, costume mishaps, and prop malfunctions require quick problem solving. Each stage is different and requires adjustments. Instant adaptations are imperative.

Being adaptable enables a choreographer to arrange, rearrange, delete, and insert movements in a work as need arises. Martha Graham would change the choreography of her roles during a premiere performance while her company scrambled to adjust. Sometimes she was in a completely different area of the stage than had been rehearsed. As a choreographer, Graham was adapting to a perceived need to revise the movement. As performers, her company members had to respond instantly and flexibly to unrehearsed changes.[29]

In her book, Twyla Tharp explained that a new work was to be based on Euripides' *Bacchae*, but during the creation of the dance, she quickly lost sight of the initial inspiration. Tharp's involvement with her dancers led her in a different direction since the state of being open and flexible allowed her to follow impulses wherever they led.[30]

Movement Explorations

Creative Possibility: (a) How many ways can you move from standing to sitting? (b) Count the number of possibilities you find. (c) Count how many ways you can move while sitting in a chair. (d) Explore your body and how it functions in relation to "chairness."

Flexibility: (a) Use the movements you developed in the previous exploration, but change their timing and facing. (b) Extend the movements using different pathways.

Adaptability: (a) Turn the chair upside down or sideways. (b) Try to repeat the same movements as in the exploration above, but adapt them to work as best as you can with the changed position of the chair. (c) Reflect on how you adapted to the new condition.

SELF-CONTROL AND SELF-DIRECTION

People who control their behavior are more likely to be focused and self-directed. Both are executive functions. Self-control inhibits inappropriate responses and turns problematic actions into those that are constructive. Self-direction promotes self-guidance and flexible responses to sensory and environmental information. Self-regulation and self-direction enable humans to persevere.[31]

Brain Network

Self-direction relates to the executive function of self-will, a function controlled by the prefrontal cortex. Injuries to this brain area produce a loss of self-determination because individuals can no longer control their actions.[32] The frontal lobes develop throughout childhood, so young children are more likely to lack control when speaking or acting, regardless of social context.[33]

Additional areas of the brain are needed in order to carry out directed actions. These include the upper gyrus, which branches from the left inferior parietal lobe. The brain produces an internal image of anticipated actions, and damage to these structures produces a disorder called apraxia, the inability to carry out skilled actions.[34]

Classroom Applications

The Montessori method is an example of self-directed learning. Students learn through their own initiative in a specifically prescribed progression and environment. Montessori teachers do not expect the students to learn from their words, but to manipulate objects in a prepared, progressive series of activities. In a Montessori science class for toddlers, students learn a principle of physics (the force of gravity) when they see a toy car roll off a table and drop to the floor, or experience botany in action when they see seeds sprout into plants when watered.[35]

Metacognition contributes to self-regulation and self-direction through self-reflection. Metacognition is high-level, critical thought that develops in maturity. It is encouraged when students engage in understanding and reflection, revealing the how and why of thinking.[36] This type of thought leads to greater self-knowledge, enabling individuals to analyze their reasoning, decision making, and planning when carrying out actions.[37]

Movement and Dance Connections

The process of molding the body into a dance instrument requires arduous work, perseverance, and self-control in both body and brain. All artistic dance forms demand practice, conditioning, and repetition to achieve dance skills. Ironically, ADHD students who have difficulty inhibiting their behaviors or directing their focus are successful in dance because it is an outlet that allows direction and release of energies. A number of studies have examined systems that can enhance performance. Mental control is improved by reducing anxiety with relaxation procedures or optimizing self-talk.[38]

Choreographers follow their impulses, but are compelled to maintain artistic intent and focus throughout completion using disciplined self-direction. They are not only driven by their own creative energies and artistic ideas but also must motivate their dancers to work to achieve their vision. Self-regulation enables a choreographer to continue to focus on an initial inspiration as it is transformed into movement and realized by the cast.

Movement Explorations

Control: (a) Do an improvisation based on creating shapes using lines in the body: straight, bent angular, and curved. (b) Select the movements that align most closely with these shapes. (c) Why did you choose those particular movements and inhibit others?

Self-Direction: (a) Choose an idea for an improvisation. (b) Design your own improvisation experience.

NOTES

1. Benjamin Bloom, *Taxonomy of Education Objectives* (New York: Longman, 1956).

2. "Anderson and Krathwohl—Bloom's Taxonomy Revised—The Second Principle," accessed November 19, 2014, http://thesecondprinciple.com/teaching-essentials/beyond-bloom-cognitive-taxonomy-revised.

3. V. S. Ramachandran, *The Tell-Tale Brain: A Neuroscientist's Quest for What Makes Us Human* (New York: W. W. Norton, 2011).

4. Torkel Klingberg, *The Learning Brain: Memory and Brain Development in Children* (Oxford, UK: Oxford University Press, 2013).

5. James P. Byrnes, *Minds, Brains, and Learning: Understanding the Psychological and Educational Relevance of Neuroscientific Research* (New York: Guilford Press, 2001).

6. Klingberg, *The Learning Brain*.

7. Ramachandran, *The Tell-Tale Brain*.

8. Sarah-Jayne Blakemore and Uta Frith, *The Learning Brain: Lessons for Education* (Malden, MA: Blackwell Publishing, 2005).

9. Noam Chomsky, *Knowledge of Language: Its Nature, Origin and Use* (New York: Praeger, 1986); Rima Faber, "The Primary Movers: Kinesthetic Learning for Primary School Children" (doctoral dissertation, American University, 1994); Klingberg, *The Learning Brain*.

10. Blakemore and Frith, *The Learning Brain*.

11. Joshua M. Langberg, Stephen P. Becker, Jeffrey N. Epstein, Aaron J. Vaughn, and Erin Girio-Herrera, "Predictors of Response and Mechanisms of Change in an Organizational Skills Intervention for Students with ADHD," *Journal of Child and Family Studies* 22 (2013): 1000–12.

12. Stephanie Harvey and Anne Goudvis, "Comprehension at the Core," *The Reading Teacher* 66, no. 6 (2013): 432–39.

13. Holk Cruse and Malte Schilling, "Getting Cognitive," in *The Neurocognition of Dance: Mind, Movement and Motor Skills*, eds. Bettina Blasing, Martin Puttke, and Thomas Schack (New York: Psychology Press, 2012), 53–74.

14. Diane L. Williams, "The Speaking Brain," in *Mind, Brain, & Education: Neuroscience Implications for the Classroom*, ed. David A. Sousa (Bloomington, IN: Solution Tree Press, 2010), 85–109.

15. Donna Coch, "Constructing a Reading Brain," in *Mind, Brain, & Education: Neuroscience Implications for the Classroom*, ed. David A. Sousa (Bloomington, IN: Solution Tree Press, 2010), 139–61.

16. Coch, "Constructing a Reading Brain."

17. Linda Elder and Richard Paul, "Critical Thinking: Intellectual Standards Essential to Reasoning Well within Every Domain of Thought," *Journal of Developmental Education* 36, no. 3 (2013): 34–35.

18. Charles M. Harris and Shenghua Zha, "Concept Mapping: A Critical Thinking Technique," *Education* 134, no. 2 (2013): 207–11.

19. Nicola Simmons and Shauna Daley, "The Arts of Thinking: Using Collage to Stimulate Scholarly Work," *Canadian Journal for the Scholarship of Teaching and Learning* 4, no. 1 (2013): 1–11.

20. Benjamin Bloom, *Taxonomy of Education Objectives* (New York: Longman, 1956); "Anderson and Krathwohl—Bloom's Taxonomy Revised—The Second Principle."

21. Brenda P. McCutchen, *Teaching Toolkits for Dance Education* (Columbia, SC: Dance Curriculum Designs, 2008).

22. H. Lynn Erickson, *Concept-Based Curriculum Instruction for the Thinking Classroom* (Thousand Oaks, CA: Corwin, 2007).

23. Larry Wright, *Critical Thinking: An Introduction to Analytical Reading and Reasoning*, 2nd ed. (New York: Oxford University Press, 2013).

24. "Critical Period," accessed April 21, 2015, http://en.wikipedia.org/wiki/Critical_period.

25. Norman Doidge, *The Brain That Changes Itself: Stories of Personal Triumph from the Frontiers of Brain Science* (New York: Penguin Books, 2007).

26. Cruse and Schilling, "Getting Cognitive."

27. Deb Curtis, Kasondra L. Brown, Lorrie Baird, and Anne M. Coughlin, "Planning Environments and Materials That Respond to Young Children's Lively Mind," *Young Children* (September 2013): 26–31.

28. Daniel Pink, *Drive* (New York: Riverhead Books, 2009); Robert Joan, "Flexible Learning as New Learning Design in Classroom Process to Promote Quality Education," *I-Manager's Journal on School Educational Technology* 9, no. 1 (2013): 37–41.

29. Bertram Ross, personal conversation with Rima Faber, summer 1960.

30. Twyla Tharp, *The Creative Habit: Learn It and Use It for Life* (New York: Simon & Schuster, 2003).

31. Glenna Batson with Margaret Wilson, *Body and Mind in Motion: Dance and Neuroscience in Conversation* (Chicago: Intellect, 2014).

32. Rita Carter, *Mapping the Mind* (Berkeley, CA: University of California Press, 2010).

33. Blakemore and Frith, *The Learning Brain*.

34. Ramachandran, *The Tell-Tale Brain*.

35. Heather Dore, "Do You Teach Them Anything? What Really Happens in a Montessori Toddler Class," *Montessori Life* (Summer 2014): 40–43.

36. Patricia L. Kolencik and Shelia A. Hillwig, *Encouraging Metacognition: Supporting Learners through Metacognitive Teaching Strategies* (New York: Peter Lang Publishing, 2011).

37. James E. Zull, *From Brain to Mind: Using Neuroscience to Guide Change in Education* (Sterling, VA: Stylus Publishing, 2011).

38. Thomas Schack, "Building Blocks and Architecture of Dance," in *The Neurocognition of Dance: Mind, Movement and Motor Skills*, eds. Bettina Blasing, Martin Puttke, and Thomas Schack (New York: Psychology Press, 2012), 11–39.

Chapter 5

Emotions

Emotions aid survival through responsive action.[1] They fuel the dance-making process and ignite audiences. Agreement is coalescing about what constitutes an emotion. The primary human emotions generally are joy, sadness, anger, fear, and surprise. They are modified to produce complex emotions such as pride or confusion. Describing emotion as feeling is misleading because conscious feeling is only part of our emotional experience.

EMOTIONS AND MEANING

Emotions are generated, stored, and connected to meaning. They are produced in response to a stimulus that is evaluated unconsciously, followed by physical responses, and culminating in conscious experiences and actions.[2] More precisely, emotions are based in the body. Emotional thoughts with their expressions and actions are found in body postures, facial expressions, and vocal tones.[3]

A recent study investigated whether facial expressions influenced self-reported emotions and physiological processes. The study focused on two muscle groups—those that produce a frown and those that move the cheeks to form a smile. It was discovered activating frown muscles made subjects feel warmer, while activating a smile generated a cooler sensation. It was theorized the temperature changes could influence the release of neurotransmitters important to emotions.[4]

Brain Network

Structures producing emotions are in the brain's limbic system—primarily the amygdala and hypothalamus (see figure 5.1). The amygdala is a central processing area for emotion, but it is more important in some emotions than others. Sometimes, activation in other areas of the limbic system minimizes the role of the amygdala.[5] Emotion begins with a stimulus—an object, situation, or recalled memory.[6] Images related to emotional stimuli from the brain's sensory centers become available from emotion-generating sites in the brain. When the amygdala receives neural patterns from these sites, it executes emotions affecting body and brain regions.[7]

The amygdala is a group of nuclei deep in the cerebral hemispheres of the brain that are thought to coordinate conscious experiences of feelings.[8] An individual becomes aware of an emotion through direct and indirect brain signals. In the direct pathway, signals are sent from the amygdala to the frontal cortex. In the indirect pathway, the amygdala sends signals to the hypothalamus, which releases hormones in the body producing muscle contractions and increased blood pressure and heart rate. These changes are

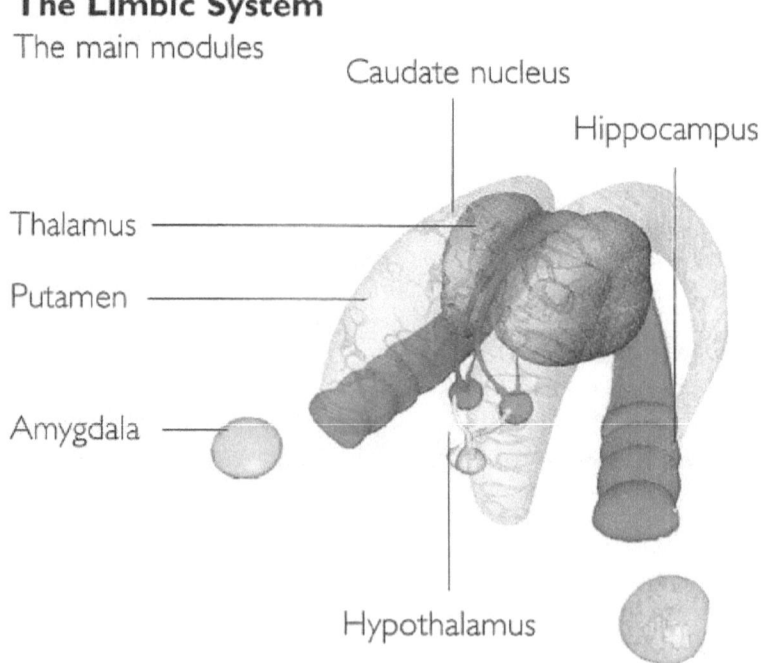

Figure 5.1   Limbic system. *Source: Mapping the Mind*, by Rita Carter. © 2010 by Rita Carter. Published by the University of California Press.

relayed to the somatosensory cortex and then to the frontal cortex, where they result in an emotion.[9]

When the amygdala is stimulated directly, fear is the strongest emotion. A neuroimaging study showed greater activity in the amygdala when subjects viewed fearful expressions compared to viewing happy faces.[10] In fearful situations, the amygdala interrupts activities and produces quick, automatic responses.[11] The amygdala is also active in response to positive stimuli. When research subjects viewed pleasant faces, input was conducted through the amygdala to the prefrontal cortex.[12] Remembered emotions are processed in the hippocampus and parts of the prefrontal cortex. Past experiences determine a person's response as well as current situations. There are a number of strong neural connections throughout the brain.[13]

In an experiment, subjects were exposed to a series of words in quick succession. A normal response to a word series is to have an attention blink (low attention), because the mind does not notice a second word if it is presented quickly after a preceding word. In the case of emotional words (*murder* or *rape*), the attention blink reaction is not present, and succeeding words gain attention because their emotional nature overrides the normal reaction to ignore them. The researcher found subjects with amygdala damage do not notice emotionally laden words under the same circumstances.[14]

Hormones known as endorphins produce positive emotions. There are many types of endorphins, with receptors in the brain and body. Although there is conflicting evidence, recent studies suggest endorphins produced in the brain contribute to feelings of well-being accompanying exercise.[15] Endorphins are among a class of chemicals that cause relaxation, opening of the body frame, and facial expressions of confidence and well-being. The molecules of these chemicals bind to specific receptors in neurons of certain brain regions to produce a natural analgesic.[16]

Anxiety and stress release a hormone called cortisol, which signals the brain and body, triggering a state of emergency. Cortisol is a life-sustaining adrenal hormone that is essential to homeostasis. It causes the metabolism to slow down during stress for economy of survival, a mechanism that has been found to produce body fat.[17]

Some authorities discuss the connection between emotions and internal sensations from the body that are mapped by the brain's insula. It is a conduit that sends and receives signals from other emotional centers—the amygdala, autonomic nervous system powered by the hypothalamus, and the orbitofrontal cortex that produces nuanced emotional judgments (see figure 5.2). Internal sensations come from the major organs of the body, bones, and kinesthetic receptors in joints, ligaments, fascia, muscles, and skin. The insula uses these sensations to assess feelings about the outside world.[18]

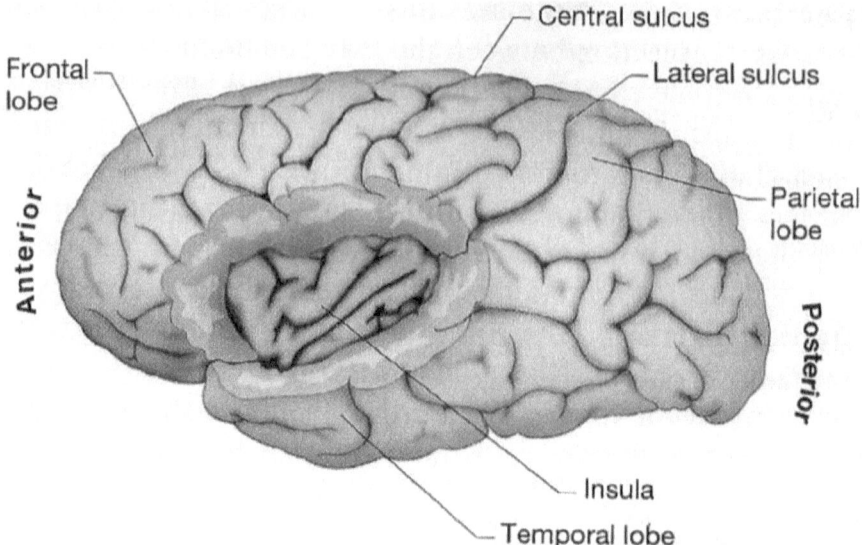

Figure 5.2 The insula deep within the brain. *Source*: Schenkman, Margaret; Bowman, James; Gisbert, Robyn; Butler, Russell, *Clinical Neuroscience for Rehabilitation*, 1st © 2013. Reproduced by permission of Pearson Education, Inc., New York, New York.

A person's emotion-recognition system determines their emotional intensity. Some people's system is highly active, resulting in great sensitivity or emotional expression. Other people have an underactive system and are slow to react to or demonstrate emotions.[19] Under or overly reactive responses may be due to psychological causes. The full spectrum of emotions is absent in forms of autism.[20]

Classroom Applications

Monitoring one's feelings and learning to channel them can be educational tools to develop self-awareness and control.[21] An effective approach is to use positive reinforcement by coupling joyful emotions and lesson content. This is accomplished by providing choices in a framework of knowledge and experiences with negotiable challenges. A challenged student is not bored, and builds confidence as long as the challenges provide balance between his or her prior knowledge and new information.[22]

Positive emotions in educational settings relate inputs to the routes traveled to or from the amygdala. In a direct emotional route, outside input goes right to the amygdala, and then is passed to the brain stem and down to the muscles, producing a reflexive response. In a less direct emotional route, input flows to the cortex before it reaches the amygdala. In the cortex, it produces a reflective, slower response modulated by data and memories. When sensory input and emotions are modulated in this way, individuals gain a deeper understanding of their experiences, leading to emotional awareness and more measured emotional responses.[23]

Movement and Dance Connections

Students frequently connect their dance making to emotional expression. Studies with students in a middle school and in an International Baccalaureate dance class previously mentioned claimed the class was an opportunity to express their emotions and gain imaginative skills—an opportunity not found in other classes.[24]

Emotions are a catalyst for art. Pioneering modern dancer Isadora Duncan was a missionary for expressive movement. She believed the body and emotions were one.[25] Duncan rejected the formality of ballet and developed a style of natural movement to communicate positive and negative emotions. Modern dancer Katherine Dunham created emotionally charged works grounded in social commentary. Her work *Southland* depicted an historical lynching of a black man accused of raping a white woman.[26] Martha Graham probed the depth of human passion and embodied expressions of grief, struggle, or anger and the open release of joy.

Performance often focuses on dances with intimate emotional content that is communicated nonverbally to an audience. Movements in a dramatic work may be abstracted from daily life, but the root of dancers' actions stem from emotional reality. Although postmodern work is often abstract with no intent of imbuing feelings, the creation of a dance often results from the choreographer's merging intellect with emotional energy.

Attention has been given to the kinesthetic response of audience members. In a 2008 survey of a Glasgow dance audience in which 40 percent of the audience had no dance training, it was deduced that people like to watch dance because it provides an emotional response along with the motor sensations experienced empathetically.[27]

Movement Explorations

Shape of Emotions: (a) Think of a negative emotion (sad, mad, afraid) and make a body shape that communicates it. (b) Reflect on the position of your body. Is your gaze outward or down? Is your spine curved or straight? Is your body closed inward or open? Are your muscles tense or relaxed? (c) Make the body shape of a positive emotion (happy, excited, proud). (d) Reflect again on the position of your body. (e) How is it different from the body shape that accompanied a negative emotion?

Movement of Emotions: (a) Extend the above exploration into movements instead of using only body shapes. (b) Find movements to express each emotion, and reflect on the types of movements performed. (c) Are the feelings evoked while you are moving and dancing?

Feelings and Social Commentary: (a) Think of situations in life. It could be at work, at school, or a social situation. (b) What emotions were experienced? (c) Express these emotions in movements. (d) Did expressing the situation in movement affect your feelings or perspective in any way about the life situation?

EMOTIONS AND LEARNING

At its most basic level, learning is an electrical and chemical process that is often tied to a student's mood or the feelings generated by the classroom atmosphere. Emotions can contribute to the level of learning accomplished or inhibit its fundamental processes. Ideas create internal brain activity. The function of emotion is to create external action in the world. When combined together, emotions serve to move ideas into action.

Brain Network

Neurotransmitters are chemicals or enzymes that facilitate the transportation of emotions. Some neurotransmitters cause excitation, making cells fire, while others shut down neural activity. Norepinephrine and dopamine are two powerful neurotransmitters that produce sensations of pleasure. Norepinephrine influences attention, perception, motivation, and arousal. Dopamine, mentioned elsewhere, is part of the learning, reward, and attention system, and is also responsible for triggering movement.[28]

Other chemicals inhibit learning. Cortisol is released during stress, making it difficult to learn new information or access memories.[29] Stress affects the adrenal glands, causing them to release cortisol and other steroids that negatively affect the amygdala and hippocampus and interfere with neurotransmitters.[30] During stress, the amygdala also becomes hyperactive and the cortex shrinks, changes that may be reversibly plastic.[31] Even imagined adverse situations can negatively affect the body by activating brain areas, including the anterior insula, which registers autonomic activity.[32]

When stress is reduced, a state of balance is restored, but a prolonged, chronic overload increases the number of activating neuronal synapses in the amygdala, leading to its hyperactivity. It also retards production of new nerve cells in the hippocampus and leads to the contraction of projections on neurons already present in the brain and shrinking of the prefrontal cortex.[33]

Classroom Applications

A positive classroom is characterized by open communication, acknowledgment of expectations, understanding and appreciation of students' efforts, student participation in decision making, and reciprocal caring of students.[34] A strong emotional connection with materials to be learned heightens the brain's receptive capacities.[35]

Some researchers believe an emotional relationship between mind and body plays a critical role in motivating learning, consciously and unconsciously.[36] The above descriptions of emotion-generating processes point to the mind/body connection. Young children are unable to modulate emotion and its physical expression due to a direct but uncontrolled reflexive link between feelings and actions. It is common for young children to cry, stamp their feet, or hit another child when provoked by a situation.

The relationship between emotions and learning was examined in a study of preschool students. A correlation was found between a child's ability to control their emotions as reported by their mother, their emotional understanding, and their cognitive understanding of lesson content. Overall interrelationships existed between emotion and cognition—a result corroborated by other researchers.[37] Emotional response to environmental cues develops before children can modify their emotions, but as they mature children learn to contain their emotions and become socialized through positive reinforcement.[38]

Researchers looked at how emotions of children, ages six to thirteen, affected learning. Youngsters viewed two films: one sad, and a second educational documentary to test learning. After viewing the first film, students were divided into four groups. One group was told the sad film was not relevant to their lives; a second group was instructed the sad film could resolve positively; a third group focused on their feelings about the sad film; and a control group had no instructions. In conclusion, the first two groups remembered more of the educational film than the other two groups. It was concluded that emotional involvement diverts attention from content.[39]

A longitudinal study on the effect of chronic stress tracked 196 American nine-to-thirteen-year-olds, half of whom spent most or all of their childhood in homes below the poverty line. Blood pressure, body mass index, cortisol, and noradrenalin levels were used as indirect measures of stress. The study showed a relationship between higher stress levels and brain function, especially with impairment of working memory. Presently, it is not known whether children recover if they experience stress during prolonged periods in childhood.[40]

Throughout the school day, students interact with each other and with their teachers. Learning occurs in a social context, and one of the most damaging experiences is to be excluded from the group.[41] How young children make sense of meeting others often predicts their ability to adjust to school and learn. Another study showed children with a more complete and flexible body of emotional knowledge often made better adaptive emotional and behavioral choices in provocative situations instigated by peers.[42]

Learning in a group setting leads to discussions of cooperation. In working collaboratively on a project, students gain from insights of others and clarify their thinking as they articulate ideas.[43] Positive cooperative learning requires careful planning, clear objectives, defined parameters, and applicable examples.[44]

Movement and Dance Connections

In the traditional four-hundred-year-old European dance pedagogy, the dance master taught holding a stick, often hitting students with it if they erred. When practices eased in the mid-twentieth century, teachers carried the stick but banged it on the floor instead. A punitive pedagogy produced fear and is still prevalent in verbal criticism. Current pedagogy is mixed. A class in which the teacher is warm and encouraging allows students to experiment with movement. A best practice is for a dance teacher to be sensitive to the emotional tone in a class.

Dance class also creates a feeling of synchrony and builds community in a group. Studies with rats found social interaction positively affects neurogenesis, or the creation of new brain cells. After twelve days of running as a group, the rats showed a significant increase in neurogenesis in comparison to rats that exercised just as much but were kept in isolation. It is possible that moving together as a group has the same effect on humans that could be coupled with heightened positive feelings.[45]

Movement Explorations

Negative Emotions and Learning: (a) Think of a dance class or group learning situation in which you were nervous or not comfortable. (b) What specific emotions could you associate with this experience? (c) Did the negative feelings affect your performance?

Positive Emotions and Learning: (a) Think of a dance class or other group learning situation in which you felt confident. (b) What specific emotions could you associate with this experience? (c) Did the positive feelings affect your performance?

Collaborative Learning: (a) Watch a YouTube video with a partner that includes instructions on how to learn a series of movements. (b) Work together with your partner to perfect the performance of the movement series. (c) What were you able to contribute to this learning situation? (d) What did your partner contribute to learning the movement series? (e) Were your contributions to learning the movement series the same or different than those contributed by your partner? (f) How did you feel about working with a partner to learn the movements?

EMPATHY AND LEARNING

Empathy is considered a sympathetic experience of feelings or thoughts of another person. Neurologically, empathy is the simulation of observed actions in the brain. As previously mentioned, this response is produced by mirror neurons that enable replication of actions, and emotions expressed by others.[46] Hence, mirror neurons facilitate learning and may be instrumental in shaping human abilities and intelligence.

Mirror neurons enable us to read the intentions of someone, or see the world from their perspective. They match the brain activity of the person observed; thus they help us know what is happening in their mind. Mirror neurons drive social integration. When a person is empathetic, similar brain cells are activated. It produces the same emotional experience as in another person.[47]

Brain Network

Mirror neurons were discovered in the premotor brain of monkeys, but in humans they seem to extend to the frontal lobe, which coordinates intention and emotion.[48] There is also an overlapping relationship between the mirror neuron system and other areas of the brain. Mirror neurons provide some insight into the processes of imitation, identification, empathy, and the possible ability to mimic the vocalizations of others.[49]

Researchers examined thirty-two subjects who viewed sad images under three different conditions: (1) watching images in a natural way; (2) watching with instructions to empathize with the images; and (3) watching while they memorized an eight-digit number along with no empathy instructions. An fMRI scan followed viewing the images. The researchers found a higher level of activity in the subjects' medial prefrontal cortex (MPFC) when empathizing with the images viewed. Activity in the MPFC was also correlated with behaviors that empathize with and help others.[50]

Classroom Applications

"Theory of mind" refers to knowing what occurs in the mind of another.[51] A group of 197 Italian fourth- and fifth-grade students were studied to learn how theory of mind was reflected in their reactions toward others. Empathy was measured with a questionnaire asking if the students could recognize other people's feelings and if they considered themselves to be the subject of emotionally laden stories. The teachers who spent the most time with the students completed a prosocial/antisocial questionnaire for each child. In conclusion, students with low theory of mind demonstrated a poor ability to understand emotions of others and to recognize inner moral states.[52]

Autism and Asperger's Syndrome are conditions in which mirror neurons may not function well. Brain-imaging research suggests there is a deficiency in the mirror neuron system of autistics, although there is some controversy about these results. The mirror neuron functioning of normal children was compared to mirror neuron functioning of a group of autistic children. Both groups were asked to observe and imitate emotional expressions while undergoing fMRI scans. The children with autism had little or no mirror neuron activity in the frontal brain area responsible for reflecting emotions.[53]

Movement and Dance Connections

Learning dance technique or choreography requires replicating demonstrated movements—an activity that relies on accurate empathy and precise operation of the mirror neuron system.[54] Sensitivity to the emotional motivation for actions can help students refine their performance. Since dance is mainly learned through replication of observed movement (see chapter 2), a positive emotional environment is paramount.

A dance performance is an experience in reciprocal empathy. Audience members empathize with the feelings expressed by the performers, but simultaneously, performers are very aware of how the audience is feeling at any point. An audience can stimulate the performers by radiating warmth and involvement, just as a lack of these qualities can sap a performer's energy.

Movement Explorations

Mirror Movement: (a) Work with a partner. One is the leader, and the other is the mirror. (b) Face each other. The leader will move in slow motion. The mirror will simultaneously copy the leader's actions. (c) Focus so strongly that you feel one another's actions.

Mirror Neurons and Body Feelings: (a) Watch a gymnastic performance on YouTube and at the same time focus on how your body feels. (b) What kind of body feelings did the performance create in your own body? (c) Did you connect any emotions with the feelings in your body?

Mirror Neurons and Empathy: (a) The next time you watch a movie, note your feelings. What is resonating for you as you watch? (b) What body movements are the actors doing, and how does your body react? What emotions are produced?

NOTES

1. Rita Carter, *Mapping the Mind*, 2nd ed. (Berkeley, CA: University of California Press, 2010).
2. Eric R. Kandel, *In Search of Memory: The Emergence of the New Science of the Mind* (New York: W. W. Norton, 2006).
3. James P. Byrnes, *Minds, Brains, and Learning: Understanding the Psychological and Educational Relevance of Neuroscientific Research* (New York: Guilford Press, 2001).
4. Tom F. Price and Carly K. Peterson, "The Emotive Neuroscience of Embodiment," *Motivation and Emotion* 36 (2012): 27–37.
5. Linda Lockwood, email message to one author, May 14, 2015.
6. Antonio Damasio, *Looking for Spinoza: Joy, Sorrow and the Feeling Brain* (Orlando, FL: A Harvest Book, 2003).
7. Damasio, *Looking for Spinoza*.
8. Kandel, *In Search of Memory*.
9. Carter, *Mapping the Mind*.
10. Byrnes, *Minds, Brains, and Learning*.
11. Sarah-Jayne Blakemore and Uta Frith, *The Learning Brain: Lessons for Education* (Malden, MA: Blackwell Publishing, 2005).
12. Judy Willis, "The Current Impact of Neuroscience," in *Mind, Brain, & Education: Neuroscience Implications for the Classroom*, ed. David A. Sousa (Bloomington, IN: Solution Tree Press, 2010), 44–66.
13. Blakemore and Frith, *The Learning Brain*.
14. Blakemore and Frith, *The Learning Brain*.
15. John J. Ratey with Eric Hagerman, *Spark: The Revolutionary New Science of Exercise and the Brain* (New York: Little, Brown and Company, 2008).
16. Damasio, *Looking for Spinoza*.
17. "Adrenal Fatigue," accessed December 1, 2014, http://www.drenalfatigue.org.
18. V. S. Ramachandran, *The Tell-Tale Brain: A Neuroscientist's Quest for What Makes Us Human* (New York: W. W. Norton, 2011).
19. Carter, *Mapping the Mind*.
20. Temple Grandin, *Thinking in Pictures: My Life with Autism* (New York: Vintage, 2006).
21. James E. Zull, *From Brain to Mind: Using Neuroscience to Guide Change in Education* (Sterling, VA: Stylus Publishing, 2011).
22. Zull, *From Brain to Mind*.
23. Zull, *From Brain to Mind*.
24. Sandra Minton, "Middle School Choreography Class: Two Parallel but Different Worlds," *Research in Dance Education* 8, no. 2 (2007): 103–21; Sandra Minton and Judi Hofmeister, "The International Baccalaureate Dance Programme: Learning

Skills for Life in the 21st Century," *Journal of Dance Education* 10, no. 3 (2010): 67–76.

25. Nancy Reynolds and Malcolm McCormick, *No Fixed Points: Dance in the Twentieth Century* (New Haven, CT: Yale University Press, 2003).

26. Reynolds and McCormick, *No Fixed Points*.

27. Corinne Jola, "Research and Choreography," in *The Neurocognition of Dance, Mind, Movement and Motor Skills*, eds. Bettina Blassing, Martin Puttke, and Thomas Schack (New York: Psychology Press, 2012), 203–34.

28. Ratey with Hagerman, *Spark*.

29. Ratey with Hagerman, *Spark*.

30. Byrnes, *Minds, Brains, and Learning*.

31. Norman Doidge, *The Brain That Changes Itself: Stories of Personal Triumph from the Frontiers of Brain Science* (New York: Penguin Books, 2007); Torkel Klingberg, *The Learning Brain: Memory and Brain Development in Children* (New York: Oxford University Press, 2013).

32. Blakemore and Frith, *The Learning Brain*.

33. Klingberg, *The Learning Brain*.

34. David A. Sousa, "How Science Met Pedagogy," in *Mind, Brain, & Education: Neuroscience Implications for the Classroom*, ed. David A. Sousa (Bloomington, IN: Solution Tree Press, 2010), 9–24.

35. Mary Helen Immordino-Yang and Matthias Faeth, "The Role of Emotion and Skilled Intuition in Learning," in *Mind, Brain, & Education: Neuroscience Implications for the Classroom*, ed. David A. Sousa (Bloomington, IN: Solution Tree Press, 2010), 69–83.

36. Mary Helen Immordino-Yang and Lesley Sylvan, "Admiration for Virtue: Neuroscientific Perspectives on a Motivating Emotion," *Contemporary Educational Psychology* 36, no. 2 (2010): 110–16.

37. A. Nayena Blankson, Marion O'Brien, Esther M. Leerkes, Stuart Marcovitch, Susan D. Calkins, and Jennifer M. Weaver, "Developmental Dynamics of Emotion and Cognition Processes in Preschoolers," *Child Development* 84, no. 1 (2013): 346–60.

38. Byrnes, *Minds, Brains, and Learning*.

39. Elizabeth L. Davis and Linda J. Levine, "Emotion Regulation Strategies That Promote Learning: Reappraisal Enhances Children's Memory for Educational Information," *Child Development* 84, no. 1 (2013): 361–74.

40. Klingberg, *The Learning Brain*.

41. Zull, *From Brain to Mind*.

42. Susanne A. Denham, Hideko H. Bassett, Erin Way, Sara Kalb, Heather Warren-Khot, and Katherine Zinsser, "How Would You Feel? What Would You Do? Development and Underpinnings of Preschoolers' Social Information Processing," *Journal of Research in Childhood Education* 28 (2014): 182–202.

43. National Research Council, *How People Learn: Brain, Mind, Experience and School* (Washington, DC: National Academy Press, 2000).

44. Linda Wilson, *Teaching 201: Traveling Beyond the Basics* (Lanham, MD: Scarecrow Education, 2004).

45. Ratey with Hagerman, *Spark*.
46. Ramachandran, *The Tell-Tale Brain*.
47. Carter, *Mapping the Mind*.
48. Carter, *Mapping the Mind*.
49. Kandel, *In Search of Memory*.
50. Lian T. Ramson, Sylvia A. Morelli, and Matthew D. Lieberman, "The Neural Correlates of Empathy: Experience, Automaticity, and Prosocial Behavior," *Journal of Cognitive Neuroscience* 24, no. 1 (2011): 235–45.
51. Carter, *Mapping the Mind*.
52. Antonia Lonigro, Fiorenzo Laghi, Roberto Baiocco, and Emma Baumgartner, "Mind Reading Skills and Empathy: Evidence for Nice and Nasty ToM Behaviours in School-Aged Children," *Journal of Child and Family Studies* 23 (2014): 581–90.
53. Carter, *Mapping the Mind*.
54. Emily S. Cross, "Building a Dance in the Human Brain," in *The Neurocognition of Dance: Mind, Movement and Motor Skills*, eds. Bettina Blasing, Martin Puttke, and Thomas Schack (New York: Psychology Press, 2012), 177–202.

Chapter 6

Memory

The phenomenon of memory has intrigued neurologists throughout the past century. Considering the importance of memory to education and the elderly, understanding memory is urgent. Baby boomers are aging, and people are living longer. There is a growing need for care concerning dementia and Alzheimer's disease. Medical advances have allowed the body to outlive brain function, but movement can aid the brain so that memory is efficient longer.

Memories provide continuity in our lives and allow us to draw comparisons between past and current experiences to envision a future.[1] They provide our sense of identity.[2] It is possible to say human memory and learning are cut from the same cloth—it is difficult to engage in one without partnering the other.

MEMORY PROCESS

Memory is the ability to absorb, store, and recall information. Some memories are experiential and relate to daily routines. Others are more conceptual and abstract, such as understanding algebra.[3] Memories are plastic. Once recorded, they can change and be rearranged by new experiences, especially experiences related to an original memory trace.[4] The dynamic nature of memory is common in court trials when witnesses are influenced by mitigating factors.

Memory is progressive and includes separate stages: encoding, storage and retrieval, or recall.[5] Encoding relates to the transformation of experiences into brain impulses. During encoding, information from the outside world reaches our senses and is combined and changed. Storage refers to the method by which the brain retains information. Retrieval raises memories into consciousness.

Brain Network

Eric Kandel's work probing the chemical processes of memory is fascinating because there are so many forms of memory, and some memories remain while others fade. Memory is the formation of an association between neurons in the brain. When one neuron in a group fires they all fire to create a pattern of activity that encompasses three stages in the formation of a memory: encoding, storage, and retrieval.[6] When a neural pattern is strong or repeatedly fired it becomes encoded and stored in the brain. Storage allows the brain to maintain information over a period of time and retrieve it in response to cues, returning information to consciousness.

On a cellular level, a new memory is formed when the synapse or gap between two brain neurons is stimulated by a thought, sensation, or perception. Repeated firing of the same neurons leaves a trace or neural pathway to create a memory.[7] The strength of communication between nerve cells at the synapse is not fixed, but it can be altered by applying different patterns of stimulation. Changes in stimulation can increase or decrease the strength of the connection.[8]

Brain structures orchestrate memories depending on the type. The hippocampus transforms encoding into explicit memories. The amygdala, central to the limbic system, regulates emotions, but as mentioned, it is more important in some emotions than others because some arise from other limbic areas.[9] The basal ganglia regulate physical activity and cognition. The cerebellum is in charge of sensate reception and motor control. Cerebral cortical areas transform neural impulses into cognition. Serotonin is a neurotransmitter that produces pleasurable feelings and is involved in memory.[10] Knowledge increases as new information is matched with memories, creating more extensive networks.[11]

Classroom Applications

Who we are stems from what we learn and remember.[12] The brain banks memories. Every experience is deposited and embeds wiring to serve as a link, but memories are malleable and change with each experience because new connections are made between nerve cells.[13] The more strongly learning is encoded, the longer it will be stored, and the more readily it is recalled. The task of education is to make memories strong, giving them value so they can be recalled but also applied in life. Herein lies the power of active learning experiences.

Simple learning is a conditioning process in which a memory is repeatedly wired until it becomes strong. This does not necessarily produce understanding. Traditional forms of learning involve memorizing facts and ideas. Higher-level learning can be achieved when content is connected with elements of previous knowledge that creates understanding. A number of factors such as timing, intensity, and consistency aid memory. Learning is most effective when students are gradually introduced to more intense stimuli.[14]

Not every experience is remembered. There are chemical memory inhibitors and suppressor genes so the mind is not cluttered with every detail of life. "Long-term synaptic facilitation requires activation of memory-enhancer genes, but also inactivation of memory-suppressor genes."[15] These inhibitors are stronger or weaker in some people, or in some areas of learning. Kandel noted the role inhibitors might play in learning disabilities has not been studied.[16]

Movement and Dance Connections

Dancers are kinesthetic learners either by nature or training. Training in many dance genres involves repetitive practice of movements not natural to the body. In ballet and most modern dance, the foot must automatically point when it leaves the ground—a motion countering the automatic toe lift in walking. This training is achieved through a multitude of movement exercises until the actions become an automatic procedural memory.

Dancers refer to muscle memory in which the body remembers movements without conscious thought. This phenomenon has not been neurologically studied, but it could occur if the memory link with the hippocampus bypasses the frontal cortex and connects directly from the cerebellum or parietal lobe. Conscious thought sometimes interferes with muscle memory, and the dancer "turns off the mind" to recall movement.

Dancers do not read their movements to perform them in the way musicians read notes. Teachers and choreographers remember movements they present, and dance students observe movements demonstrated in class and rehearsals. Learning movement sequences immediately is required at auditions, and memorizing movement quickly is a skill learned as an aspect of dance training.

All three stages of memory are involved when learning movement. During encoding, dancers observe and then fit movements to their own body. The movements are transferred by electrical impulse patterns sent to the brain. During storage, movement impulses are deposited in memory. Later, the same movement impulses are recalled during class, rehearsals, and performances.

Movement Explorations

Encoding: (a) Think of a routine movement activity that you perform on a daily basis. An example is getting dressed. (b) Perform the same routine set of movements without the objects (clothes). (c) Think about each aspect of your movements (lifting and extending your arm to put it in a sleeve). (d) When you do each of these activities daily, are you conscious of each of your movements, or have they become automatic?

Storage: (a) Create a short sequence of movements. (b) Practice the movement sequence until you believe you have committed it to memory. The movement sequence should be committed to memory if you can perform it readily without giving it a great deal of thought.

Retrieval: (a) Test your memory of the movement sequence you created in the previous exploration by performing it several days later. (b) Were there any cues or ideas that helped you remember the movement sequence?

SHORT-TERM MEMORY

Memory interplays between two systems: short-term and long-term memory. As the term implies, short-term memory retains information for only a short period of time. After that, synaptic connections weaken and memories fade if not passed to long-term memory.

Short-term memory holds information in the mind while working on a problem.[17] Also called working memory, studies show it increases throughout childhood, extending into the teenage years. The same studies demonstrate there is great variation in the short-term memory capacity of children from the same age group, plus there is high correlation between visuospatial working memory and mathematical skills and between verbal working memory and reading abilities.[18]

Brain Network

On a cellular level, short-term memory can be strengthened, weakened, or changed. Weakening occurs when connections are not reinforced through repetition or changed by new experiences. The strengthening of synaptic connections during short-term memory is produced through the release of a series of chemicals—serotonin, AMP (a small molecule regulating signaling between cells), and kinase—which leads to releasing the neurotransmitter glutamate to provide a pathway for short-term memory connections.[19]

A brain model of short-term memory includes three parts: (1) the central executive that coordinates information from various sources, organizing and switching attention as needed; (2) the visuo-spatial sketch pad that holds images; and (3) the phonological loop that holds acoustic and speech information.[20]

Research on memory systems revealed there are two components to short-term memory. Initially, there is a brain area that processes incoming information for a few seconds, called immediate memory. A second brain function, working memory, consciously processes information for a more extended period, but this information is discarded if it no longer serves a purpose. The brain must make sense of information or its meaning to commit information to long-term memory.[21]

Classroom Applications

The initial goal in education is to achieve focused attention that captures short-term or working memory. The extent to which working memory is engaged determines student focus and concentration. The brain is attracted to new or novel experiences; therefore, teachers must provide balance between familiar and new information. Students who exhibit concentration issues need few distractions so learning is not inhibited by constant intensity. Research revealed that subjects with greater working memory capacity maintain their focus longer, while those with smaller capacity are distracted more easily.[22]

Learning is facilitated through manipulation of information in working memory. Experiential learning connects momentary thoughts to create continuity and meaning. It helps students rearrange input in order to solve problems. One of the reasons experiential learning is so effective is it combines visuospatial and sensory information, reinforcing facts with sensory cues.

Researchers examined the updating of working memory in sixty-three monolingual English-speaking children, K–2. Four of the eight tasks in the study assessed the ability to update working memory used to monitor, manipulate content, and track characters in a story. As a result, the capacity to update working memory was related to remembering characters over multiple exposures.[23]

Movement and Dance Connections

Dancers use short-term memory when first learning movement since there is a need to hold and manipulate individual actions in mind before they are connected to form sequences and whole dances. Training for the ability to pick movement up quickly exercises short-term memory. If choreography is not reinforced by repetition, neural connections weaken and movements are forgotten.

In presenting a movement sequence, a teacher or choreographer will start with an initial short series, and build on the series in chunks, one phrase at a time. The new segment will be practiced alone, and then added to the phrase being built. In this way, dancers can learn long movement sequences quickly and increase their working memory capacity.

Choreographers often improvise to spontaneously create movement possibilities. Working memory is used to recall improvised movements, decide which movements to use, organize them, and create sequences to fulfill choreographic intent. Video is an excellent tool to track improvisational ideas, especially for choreographers who cannot readily remember improvised movement.

Movement Explorations

Super Short-Term Memory: (a) Watch a YouTube video of a short dance only once. (b) Try to perform as much of the dance as you can remember. (c) Note which aspects attract your attention and remain in your memory.

Working Memory: (a) Perform each of the following movements in sequence one after the other. Step to the side, run in a circle, hop in any direction, leap three times, and balance for five counts on one foot. (b) Once you have performed the movement sequence, rearrange their order in your mind, and perform this new sequence. (c) Was it difficult or easy for you to mentally rearrange the movements and perform them in the new order?

Visuospatial Memory Connections: (a) Using the same movement sequence you used in the previous exercise, mentally travel or place each of the movements in a different direction or location in space. (b) See if you can remember where you set each action. (c) Perform each of the movements in the sequence in their new directions or locations. (d) Did the connection between the direction or location of a movement help your memory?

LONG-TERM MEMORY

Long-term memory represents consciousness of the self and provides the thread that ties together events in our lives. It stores and retains all the information and experiences we acquire through study and from personal experiences. It endures.[24]

There are structural differences between short-term and long-term memory. Short-term memory is transformed into long-term memory actively through repetition or passively through sleep. The frequency of repetition or the intensity of original stimuli determines which memories become long term. As mentioned, long-term memories can change over time. The brain is plastic, and memories are colored by or open to change from new experiences or information.

Brain Network

Long-term memories are initiated in the hippocampus—more specifically in the pyramidal cells of the hippocampus.[25] Consolidation, the process of laying down long-term memories, involves a replay between the hippocampus, the cortex, and back again. It turns fleeting impressions into long-term memories so they are embedded or etched into cortical tissue. It also links memories together related to the same experience.[26]

Different types of long-term memories are associated with different parts of the brain and distributed throughout interconnected neural networks.[27] The hippocampus redirects memories for storage in different parts of the brain. Emotional memories are stored in the amygdala, language memories in the temporal lobe, visual information in the occipital lobe, and sense of touch and movement in the parietal lobe.[28]

Declarative or explicit and procedural or implicit are two types of long-term memory. Declarative/explicit memory involves selective attention during encoding, recalls words and events, and organizes them.[29] Doctors discovered the function of the hippocampus in declarative/explicit memory from treating an epileptic individual by removing his hippocampus. It became impossible for the patient to remember new information. From this it was deduced that the hippocampus transforms short-term into long-term memories.[30]

Procedural/implicit memory refers to remembering how to automatically perform an action or task;[31] for example, riding a bicycle or brushing your teeth. These memories, are formed through habits, sensitivities, classical conditioning, and as perceptual motor skills[32] that are remembered but hard to explain because they are stored in our reflexive, unconscious mind.[33] Nonverbal interactions with people and emotionally based memories are also part of this system; in the first years of life, children rely on procedural memory to learn basic skills.[34]

Memory is not a single system because it relies on several functions in structures deep in the cerebral cortex, including emotional associations in the amygdala, habits involving the striatum (the caudate nucleus and putamen together), and the cerebellum to learn motor skills. The individual, described previously, who had his hippocampus removed could still learn automatic skills and remember those acquired previous to his surgery.[35]

Integration and resonance of experiences with what is already known are primary to long-term memory or learning. Effort expended during the learning process is also a factor, as is spending more time in working memory. Effort combined with trial and error may be uncomfortable, but demonstrated learning is occurring.[36] As mentioned, stress interferes with the formation of memories because it produces glucocorticoid that kills hippocampal cells.[37]

To explain further, long-term potentiation occurs when a thought or behavior is repeated and produces a chronic stimulation of the same pathway in the brain. When that happens, glutamate stimulates both AMPA and N-methyl-D-aspartate (NMDA) receptors. NMDA receptors are normally blocked by magnesium but become unblocked when the stimulation is "fast and furious" and opens the NMDA receptors. It acts to strengthen the connection between the two cells so the pathway is more sensitive with future stimulation.[38]

Classroom Applications

Information about human memory endorses novel approaches for improving student learning. One strategy is to integrate literature, the arts, and movement in lessons, especially when teaching young children. Authorities believe this approach works because literature and the arts are based on symbol systems such as images, words, and gestures that convey meaning and are strong stimuli.[39] Integrated arts is also successful with children that have learning issues because learning through different sensory modalities taps into memory in different areas of the brain and circumvents areas of dysfunction.

In a study, adult education professionals and college students created drawings highlighting relationships in their experiences. It was concluded the drawings supported memory and reflection.[40] In another study, researchers investigated active teaching strategies' effect on seventh-graders' long-term memory in comparison to copying scientific terms. Active learning included talking to classmates about the meaning of science terms and drawing pictures that represented their meaning. When the students were quizzed on the science terms, use of active teaching strategies increased word retention, especially in struggling readers.[41]

Laboratory-based research demonstrated there are long-term memory benefits to studying in multiple or spaced sessions rather than in one long period. However, under realistic conditions, although students were aware of the benefits of spaced study, they were more likely to use this strategy when a great amount of material was to be learned, the exam had great weight, the exam was difficult, or the material to be learned was greatly valued.[42]

Movement and Dance Connections

Procedural/implicit memory is particularly relevant in dance to attain motor skills and techniques. Ballet, modern dance, and jazz students condition their bodies to automatically point their toes and turn out the leg from the hip without conscious thought so they can concentrate on performance movements.

Declarative/explicit memory comes into play when learning new or complicated skills requiring conscious thought. When recalling a work, declarative/explicit memory functions to analyze or describe movements.

Remembering dance synthesizes a variety of sensory modalities: proprioceptive, spatial, visual, temporal, and auditory. Dance can stimulate long-term kinesthetic memories for both dancers and audience members. From an audience perspective, dance also activates visual and auditory musical or sound memories.

Improving memory for movement is a necessity for the professional. Long phrases are replicated quickly in auditions or classes. Movements connected to prior knowledge are remembered more easily, as are movements presented in spaced practice sessions such as rehearsals. It is interesting that dancers who have more difficulty learning sometimes make better teachers; their struggle implants movement in their minds in greater detail. Accompaniment and other sounds also aid movement memory and serve as cues. Mind maps or drawings enhance memory of choreography.

Movement Explorations

Long-Term Memory: (a) Try to remember the movement sequence you performed in the short-term memory section without referring back to it. (b) If you remembered it, what was your process in recalling it?

Procedural/Implicit Memory: (a) Think of movements you perform to walk. (b) Analyze your actions. (E.g.: pushing off with your back foot, lifting your leg, shifting your weight.) (c) Select a movement and describe it in detail. (d) Was it easy or difficult to break the action apart into specific components? Why do you think this was so?

Declarative/Explicit Memory: (a) Think of a movement memory and accompanying physical sensation from a past action. (b) Take some time to review it. Do you feel the memory in your body? Do you have associations or connections to the feeling?

CONSCIOUSNESS

Consciousness is awareness of the self in relation to sensations, motivations, and cognition of reality. It is the sense of "me" as a cognizant and mindful individual. It is believed that humans have the highest, most developed level of consciousness compared to other animals. Humans have metacognition or the ability to analyze and regulate their thinking.

Consciousness is a mental model of the world created using memory feedback loops in different parameters such as temperature, space, time, or the self in relation to others. Sensory input and experiential stimuli create awareness of the present, but a continuum of life experiences is developed through memory. Humans conceptualize time by remembering the past in order to project into the future.[43]

Sigmund Freud built his theory of psychoanalysis on recognition of the power of the unconscious. He envisioned the mind akin to an iceberg. Some 10 percent of an iceberg floats above the water (conscious mind), and 90 percent is below the surface (unconscious). This does not render 90 percent dormant since thoughts and behaviors are affected by the unconscious collective of life experiences. Although Freud's theories are now generally discredited, an understanding endures that the unconscious mind is a great motivator of behavior.[44]

Brain Network

Memories are neurologically categorized based on their function. Consciousness is subdivided into semantic and episodic functions. Semantic memories are those of objects and historical or cultural events. Episodic memories record personal events and experiences much like a diary.[45]

Neurologists use brain imaging to investigate states of consciousness and unconsciousness. Consciousness is characterized by a person's observable behaviors correlated with neuronal biophysical information. After behaviors are used to identify consciousness, it is possible to identify the brain mechanisms characteristic of that state. As voiced previously, while it is possible to correlate brain activity with thoughts and behaviors, we cannot explain human actions or meaning through brain functions; consciousness is not understood through physiology.[46]

Classroom Applications

Traditional education focuses on semantic memory. Information is memorized, and learning is accomplished through books. Rousseau challenged this approach by promoting learning through nature. John Dewey developed a pedagogy based on learning through experiences by combining sematic with episodic memory to create a complete learning environment.

Until learning is either experienced or made "real," it has no meaning. Rote learning does not develop understanding or engage the learner. Consciousness is an awareness of the self in relation to content and gives content relevance. Episodic memory is the collection of events and thoughts creating a continuum of life. Semantic learning expands awareness beyond the personal to broaden scope and depth of understanding. We learn about objects and events to enrich life experiences and avoid past mistakes. History and science are built on the union of semantic and episodic memory through human consciousness—an understanding of the past to forge a future.

Movement and Dance Connections

Dance is episodic. The body is personal, present, and experiential. The body, in coordination with the brain, holds a collection of life experiences and events that are reflected through its shape and movements. Martha Graham said, "The body doesn't lie." She also said "What you do shapes your body." People who understand the body and body language can read a person's body and understand them through their body use and movement quality. The body is a physical manifestation of a person's consciousness.

Due to intense technical training, dancers often neglect semantic learning about the history of dance, or about dance as art. The growth of dance in higher education broadened the scope of dance education by including historical continuity and cultural significance. When this learning is combined with episodic body wisdom, it produces knowledgeable depth and movement skill.

Over the years, choreographers develop a style made up of movements they have used repeatedly. These movements have a special use or meaning for the choreographer and function like a vocabulary or dictionary of semantic movement memories. Episodic memories can be a jumping-off point for creating movements, or help a dancer remember specific dance experiences from the past.

Movement Explorations

Semantic Memory: (a) Select a description of an important historical event from a book or the Internet. (b) Read this description carefully followed by selecting three or four aspects of the event that resonate with your understanding of the overall description. (c) Create and perform a simple movement that you feel reflects the meaning of each aspect of the event. (d) At a later time, perform the movements you created, and at the same time recall the meaning of each aspect. (e) Did performing the movements help you remember the meaning of aspects of the historical event?

Episodic Memory: (a) Select an event from your past and choose three or four aspects of the event that resonate for you. (b) Create and perform a simple movement that you think reflects the meaning or feeling that accompanied each aspect of the event. (c) A day or two later, perform the movements again, and at the same time recall the meaning of each aspect. Do you think performing the movements helped you remember the meaning of aspects of the event? Did understanding the meaning help you perform? Did movement change the memory?

SPATIAL MEMORY

Humans continually map the spatial relationship in the surrounding environment and commit it to memory, an ability with adaptive evolutionary and prehistoric roots. Before the advent of writing, the most successful humans hunted or gathered well by creating spatial mind maps using nonlinear visualization. Written language is linear, and success in the civilized world required abstract, logical, linear thought. The explosion of technology, networking, and computerized imagery returns to a spatial and visual nonlinear orientation, and success necessitates holistic brain patterns.[47]

Brain Network

The human vestibular apparatus provides balance and orientation in space. This apparatus consists of three semicircular canals in the inner ear that organizes perceptions oriented according to gravity and three egocentric dimensions of space—vertical, horizontal, and front/back. They inform a sense of position and motion in three-dimensional space[48] and function in coordination with visual perception (see figure 6.1).

Many animals, including humans, develop mind maps of their surroundings and remember locations in space through different brain structures. When seeing something located diagonally up to the right, certain neurons in the intraparietal cortex record the position, but different neurons are active when observing an object located diagonally up to the left. A map also exists in the frontal lobe coordinating information from the area controlling eye movements and tracking locations. Other parts of the brain record visual input like colors or motion.[49]

The hippocampus does not export long-term memories of space for cortical storage. Spatial memories remain encoded in the pyramidal or place cells of the hippocampus. The brain combines sensory modalities to produce memories of spatial location by relying on input from visual, auditory, and tactile information, and records spatial locations as the body travels in space. The role of the hippocampus in spatial memory was confirmed in a study of London cab drivers who undergo required intensive vocational training. Their hippocampi were active as they described a route followed throughout the city and became enlarged from their training.[50]

Classroom Applications

Geometry partners with spatial memory. Kindergartners learn geometric shapes and are expected to recognize basic design. Second-graders are trained to mentally move geometric shapes by flipping, turning, or sliding them, while third-graders begin to assess measurements. Geometric operations require recall of spatial memories.[51]

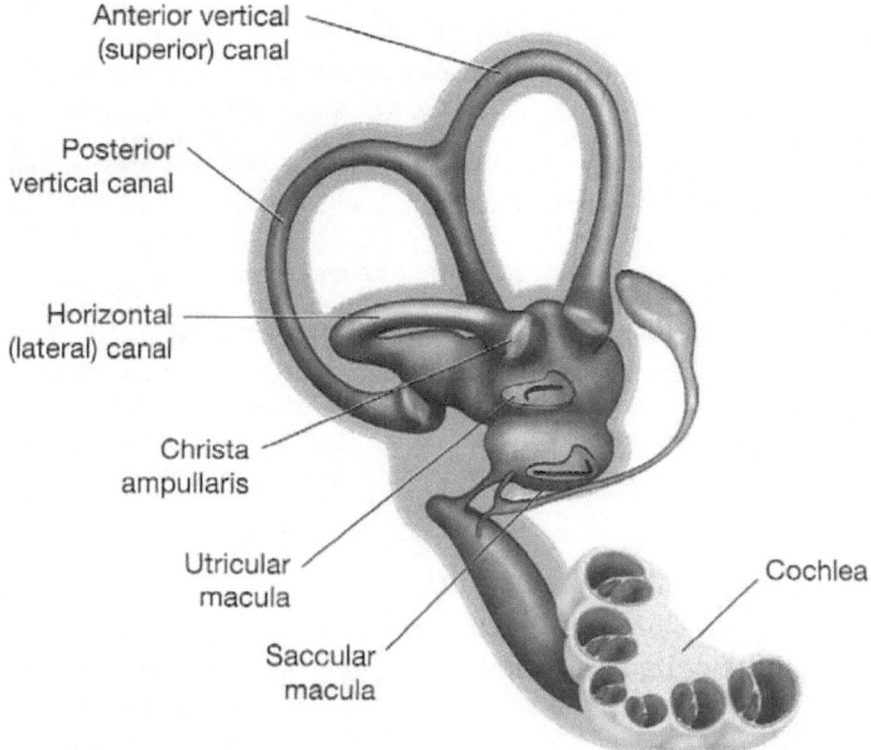

Figure 6.1 Vestibular apparatus in inner ear. *Source*: Schenkman, Margaret; Bowman, James; Gisbert, Robyn; Butler, Russell, *Clinical Neuroscience for Rehabilitation*, 1st © 2013. Reproduced by permission of Pearson Education, Inc., New York, New York.

Spatial brain maps are based on a purpose: spatial reasoning, predicting future movements, or remembering the location of objects or places.[52] Brain maps are either egocentric— location of objects based on the position of an individual—or allocentric—recorded based on position relative to the environment, or how objects relate to one another.[53] The location of a bedside lamp in relation to a person lying in a bed is egocentric, but the lamp's position in relation to the room's door is allocentric. If a person turns right or left, the change is egocentric, but traveling north or south is allocentric.

Movement and Dance Connections

Spatial components in dance include direction, level, size, shape, pathway, and position. Dancers perform movements of varying sizes, in many directions, on different levels, in a variety of pathways, and with body shape changes.

Dancers need to know their relative and constantly changing spatial relationships to other dancers in relation to the studio or stage while seeking instantaneous orientation to gravity. During class or performance, dancers use both egocentric and allocentric mind maps: egocentric relationships between dancers to prevent collisions; allocentric in relation to the stage and environment.

Dancers develop spatial schema in relation to their bodies. Whether intuitively applied or formally defined, the body is used in three-dimensional planes. These are categorized as: traverse or horizontal; sagittal, dividing the body right and left vertically; and coronal, dividing the body front and back vertically. Movements that function diagonally across the body's median are translateral and involve both hemispheres of the brain. Diagonals are neurologically more complex spatial patterns than same side or forward and backward actions.

Movement Explorations

Spatial Orientation: (a) How could you move your whole body to explore each of the three egocentric orientations: vertical, horizontal, forward/backward? (b) Close your eyes and do it again. (c) Can you retain your orientation?

Mind Mapping: (a) Read instructions for the following movement sequence and create a map in your mind of it. (b) Walk four steps forward, take one step to the right side, look to your left, jump, turn, and skip. (c) Perform the movement sequence without reading it. (d) Could you remember the movement sequence? (e) Did you recall the map in your mind?

Spatial Accuracy: (a) Stand at a distance from a wall that you imagine to be the length of your arms. (b) Stretch out your arms to see if you can touch the wall. (c) Do your fingertips touch the wall? (d) Were you accurate in judging the correct distance from the wall?

Egocentric Spatial Relationships: (a) Select three objects in a room and describe where in space they are positioned in relation to your body. For example, from where I am seated there is a lamp up high and on a diagonal in relation to my body. (b) Change your position in the room and again describe where in space the same three objects are now located. (c) Were the three objects oriented spatially in relation to your body in the same way or in a different way this second time?

Allocentric Spatial Relationships: (a) Think of where the three objects are placed in the room. (b) Stand at the door. Close your eyes and walk to each object. (c) Were you able to do it? (d) What was in your mind to accomplish the task?

THE EMOTION/MEMORY LINK

In an experiment, subjects were asked to think about one emotional episode in their lives that was powerful and involved happiness, sadness, fear, or anger by focusing on the details and imagery associated with the event. It was found that adults who experienced emotionally laden events are almost able to re-experience the associated emotions.[54] Emotions can color and distort memories. Conversely, it has been well noted that extremely traumatic experiences can produce amnesia.

Brain Network

Long-term emotional memories are stored in the amygdala. In general, negative memories produce avoidance of events and a tendency to seek situations that generate positive feelings.[55] The intensity of an experience or level of wakefulness are critical to experiencing accompanying feelings.[56] Traumatic, stressful events can cause flashbulb memories, strong flashes of memory like the post-traumatic stress experienced by veterans. When the hippocampus recalls such negative, vivid memories, adrenaline and noradrenaline are released in the bloodstream to produce a fight-or-flight reaction.[57]

There is a correlation between emotions, memory, and body states. Changes in the brain initiate automatic physiological reactions accompanying crucial mental experiences. The hippocampus must send messages to brain areas that control physiological experiences for storage and recall. Threats to individuals and subsequent reactions are mapped in the central nervous system as neural maps that trigger corrective responses and interrupt those responses once an issue is rectified.[58]

Classroom Applications

Children's emotions are immediate and intense. From the moment of birth they are building an emotional memory bank. By the time they enter the classroom, their early environment has shaped innate, genetic personalities with attitudes, motivations, reactions, and emotions toward classroom learning.

Students' memory and learning can be heightened through the positive aspects of play, particularly in primary students. Research shows play is important to the development of brain functions when combined with appropriate teaching strategies, and can be adapted to different types of learners.[59]

An experiment tested the hypothesis that negative arousal leads to memory narrowing, while positive arousal broadens recall. Seventy-two college students observed an equal number of positive and negative pictures on a computer screen and were instructed to study them. The results of later memory trials showed positive images led to an increase in recognition of both central and peripheral picture details. Negative images caused students to focus on central details and less on those in the periphery.[60]

Movement and Dance Connections

Movement erupts when emotions are too strong to contain in words.[61] Dance is a personal and cultural expression. In the era of Impressionism, François Delsarte contended that body postures and the position of body parts could be codified and reproduced to communicate emotions (see figure 6.2). Isadora Duncan's choreography was an embodiment of Delsartian ideals.

The field of dance therapy uses physical expression of emotional memories to help people understand and resolve issues. Dance therapists discover clues to help people by studying their actions.[62] Some psychiatrists assert a person experiences the world only through the body.[63] Movement is integral to Gestalt therapy, an expressive therapeutic dynamic.[64]

Movement Explorations

Emotion, Body States, and Memory: (a) Sit in a quiet spot and focus on a memory with emotional content from your past. (b) Recall the memory with all of your senses. (c) Assess how your body feels as you review this memory. (d) Does your body react to the emotions?

Positive Memories: (a) Bring an important positive memory to mind. (b) Recall the memory with all of your senses. (c) Does your body react? (d) Transform your feelings into movement.

Negative Memories: (a) Bring an important negative memory to mind. (b) Recall the memory with all of your senses. (c) Does your body react? (d) Transform your feelings into movement.

Flashbulb Memory: (a) Focus on a memory that is strong and that springs to mind instantly. (b) How would you describe the feelings you have associated with this memory? (c) Does your body react to the emotions? (d) Transform your feeling response into movement.

RECALL

Memories are not conscious until they can be recalled. Cues must relate to an original experience. They can be external stimuli experienced as sensations or internal triggers like ideas or urges.[65] One of the most powerful recall cues is the emotional context of a memory.[66] Negative, painful memories can interfere with recall and discourage learning.[67]

Figure 6.2 The pose pictured communicates an emotion. *Source*: Reprinted by permission, from S. C. Minton, 2007, *Choreography: A Basic Approach Using Improvisation*, 3rd ed. (Champaign, IL: Human Kinetics). Photographer Joe Clithero of B & J Creative Photography.

Chunking is a memory strategy related to the structure and function of information to be recalled. The chunks are sometimes called schemas. Chucking is the grouping of connected memories together. Experts in a field are more likely to use chunking to remember and recall content than novices.[68]

Researchers matched twenty-one actors with twenty-four controls who had no acting training. What differentiated the groups was the actors' ability to extract and recall the gist of a story from verbal content, chunking information related to a central theme.[69]

Brain Network

Recall in the brain reactivates the areas that were stimulated in an original experience. The more intense an experience, the more easily the recall system is re-stimulated, except in extreme situations of repression or amnesia. The ability to recall information is dependent on a distributed network of brain regions and activation of Broca's area, the prefrontal cortex, globus pallidus (part of the basal ganglia), anterior cingulate cortex, thalamus, and cerebellum.[70] These areas facilitate reconsolidation.

The neural reconsolidation process functions to check memories and, if applicable, they become stronger. Recent research indicates reconsolidation is related to the generation of new neurons, especially in the hippocampus.[71] Reconsolidation can also alter memories since memories are plastic; each time a memory is recalled it can change, and the slightly altered memory replaces the previous one.[72] Memory retrieval may also be affected by acute stress, which can chemically inhibit memory formation.[73]

Classroom Applications

The process of retrieving memories can be simple or complex. Recognition is simple retrieval triggered when we recognize something in the present. Complex retrieval is based on the connection between ideas and their application in the real world. Learning strategies that include performance or application of information enable students to retrieve information and generate new knowledge.[74]

Cues that work have neuronal firing patterns that overlap with an aspect of the original neural pattern as it was laid down. Helpful recall cues include categories, places, subjects, people, and times that relate to a memory. An intense or consistent experience provides strong recall cues, hence practice is effective.[75] Recall cues are also provided in positive meaningful materials.[76] Play, as a teaching strategy, stresses meaningful fun with positive feelings that aid recall.

Mneumonics is a system of cueing memory based on creating associative bonds by partnering information with words, images, or facts. For example, the thick and elastic nature of arteries can be remembered easily using the mneumonic "Art (ery) was *thick* around the middle so he wore pants with an *elastic* waistband."[77]

Cues that reinforce one another in abundance strengthen recall. When memories are stored in several brain locations, cues can be stimulated from a broader set of connections. In reverse, recalled memories can become fragile and weak. Extinction takes place when cues used to recall a memory are not reinforced, inconsistent, or contradict one another.

Movement and Dance Connections

Dancers recall movement with great individuality depending on learning style. They cue from spatial location, movement shapes, musical cues or counts, movement meaning, and relationships between dancers. Other recall techniques include bodily sensations generated when performing a movement, the artistic intent or emotional content communicated, or the organization of a movement sequence.

Repetitive practice has great value in movement recall. Repeating movements generates numerous impulses over the same neural pathways, aiding memory consolidation. Learning material at the last minute does not create well-formed patterns, can be stressful, and does not provide time for memory consolidation.

Inquiry-based learning, a form of active learning, is being introduced in dance classrooms to aid recall. Learning movement through analysis can aid memory retrieval by connecting movements to specific cues in multiple areas of the brain. It produces understanding of the artistic intent and structure of the movement, and it enriches memory.

Chunking movement into groups constructs patterns or phrases. The adage "whatever is wired together is fired together" holds true, because remembering part of a pattern may evoke its entirety. Chunking also permits choreographers and dancers to create a sense of artistry. Groups of movements that repeat form patterns, holding movements together to create aesthetic phrases.

Movement Explorations

Visual Recall: (a) Find a short dance on YouTube and select one movement sequence from this dance. (b) Practice the selected movement sequence in front of a mirror so you can analyze its design. (c) Wait a week and recall the movement sequence by focusing on its visual aspects. (d) Did recalling the visual aspects of the movements help you remember?

Body Feeling and Recall: (a) Use the same dance or find a new one and select a different movement sequence. (b) Practice this movement sequence while focusing on how it feels in your body. Notice changes of sensations in your body while performing the sequence. (c) Wait a week and recall the movement sequence by focusing on the body. (d) How would you compare the sensate method of movement recall with the visual method?

Spatial Cues: (a) Create a movement sequence in which you move throughout the space. (b) Practice your movement sequence while focusing on the spatial area in which each action is performed. (c) Wait a week and recall your movement sequence while focusing on the spatial area connected with each action. (d) Did remembering the location of each action in the sequence help your recall?

Relationship Cues: (a) Create a movement sequence with a partner. (b) As you and your partner practice the sequence, pay careful attention to how you both relate. (c) Wait a week and perform the movement sequence again with your partner. (d) Did recalling your relationships with your partner help your recall?

Emotional Cues: (a) Create a movement sequence that is based on a particular emotion. (b) Practice the movement sequence while focusing on the feeling. (c) Wait a week and recall the same movement sequence. (d) Was it easy or difficult for you to recall the movement sequence?

Chunking and Recall: (a) Learn a movement sequence by focusing on its structure and organization. (b) Wait a week and perform the sequence again by recalling its structural organization. (c) How would you compare your ability to recall movements you created in comparison to recalling a movement sequence created by someone else?

NOTES

1. Eric R. Kandel, *In Search of Memory: The Emergence of a New Science of Mind* (New York: W. W. Norton & Company, 2006); David Eagleman, *The Brain: The Story of You* (London: Canongate Books, 2015).

2. Nancy Andreasen, *The Creative Brain* (New York: Penguin Group, 2005).

3. Kandel, *In Search of Memory*.

4. Norman Doige, *The Brain That Changes Itself: Stories of Personal Triumph from the Frontiers of Brain Science* (New York: Penguin Books, 2007).

5. "Memory," accessed June 6, 2015, http://en.wikipedia.org/wiki/Memory.

6. Rita Carter, *Mapping the Mind* (Berkeley, CA: University of California Press, 2010).

7. Lynn Helding, "Memory, Hither Come," *Journal of Singing* 69, no. 3 (2013): 337–44.

8. Kandel, *In Search of Memory*.

9. Linda Lockwood, email message to one author, May 14, 2015.

10. Patricia Wolfe, *Brain Matters: Translating Research into Classroom Practice* (Alexandria, VA: Association for Supervision and Curriculum Development, 2001).

11. Judy Willis, "The Current Impact of Neuroscience," in *Mind, Brain, & Education: Neuroscience Implications for the Classroom*, ed. David A. Sousa (Bloomington, IN: Solution Tree Press, 2010), 44–66.

12. Eric Kandel, "The Molecular Biology of Memory Storage: A Dialogue between Genes and Synapses" (keynote lecture, Learning and the Brain Symposium, Washington, DC, April 7, 2011).

13. Kandel, *In Search of Memory*.

14. James E. Zull, *From Brain to Mind: Using Neuroscience to Guide Change in Education* (Sterling, VA: Stylus Publishing, 2011).

15. Kandel, *In Search of Memory*, 55.

16. Eric Kandel in conversation with Rima Faber, Learning and the Brain Conference, New York City, April 7, 2011.

17. "Memory," accessed June 9, 2015, http://en.wikipedia.org/wiki/Memory.

18. Torkel Klingberg, *The Learning Brain: Memory and Brain Development in Children* (New York: Oxford University Press, 2013).

19. Kandel, *In Search of Memory*.

20. Carter, *Mapping the Mind*.

21. David A. Sousa, "How Science Met Pedagogy," in *Mind, Brain, & Education: Neuroscience Implications for the Classroom*, ed. David A. Sousa (Bloomington, IN: Solution Tree Press, 2010), 9–24.

22. Klingberg, *The Learning Brain*.

23. Cristy Whitely and Paola Colozzo, "Who's Who? Memory Updating and Character Reference in Children's Narratives," *Journal of Speech, Language, and Hearing Research* 56 (2013): 1625–36.

24. Klingberg, *The Learning Brain*.

25. Kandel, *In Search of Memory*.

26. Carter, *Mapping the Mind*.

27. Helding, "Memory, Hither Come."

28. Michio Kaku, *The Future of the Mind: The Scientific Quest to Understand, Enhance, and Empower the Mind* (New York: Doubleday, 2014).

29. Helding, "Memory, Hither Come."

30. Doige, *The Brain That Changes Itself*.

31. Helding, "Memory, Hither Come."

32. Kandel, *In Search of Memory*.

33. Doige, *The Brain That Changes Itself*.

34. Doige, *The Brain That Changes Itself*.

35. Kandel, *In Search of Memory*.

36. Helding, "Memory, Hither Come."

37. Doige, *The Brain That Changes Itself*.

38. Linda Lockwood, email message to one author, May 27, 2015.

39. Karyn W. Tunks and Rebecca M. Giles, "Read Aloud, Sing Along, and Move Around: Musically Motivating Children with Books," *Perspectives* 8, no. 3 (2013): 6–12.

40. Julia Ellis, Randy Hetherington, Meridith Lovell, Janet McConaghy, and Melody Viczko, "Draw Me a Picture, Tell Me a Story: Evoking Memory and Supporting Analysis through Pre-Interview Drawing Activities," *Alberta Journal of Educational Research* 58, no. 4 (2013): 488–508.

41. Rebecca Shore, Jenna Ray, and Paula Goolkasian, "Too Close for (Brain) Comfort: Improving Science Vocabulary Learning in the Middle Grades," *Middle School Journal* (May 2013): 16–21.

42. Jonathan A. Susser and Jennifer McCabe, "From the Lab to the Dorm Room: Metacognitive Awareness and Use of Spaced Study," *Instructional Science* 41 (2013): 345–63.

43. Kaku, *The Future of the Mind*.

44. Ernest Jones, *Life and Work of Sigmund Freud, Volume I* (New York: Basic Books, Inc., 1953).

45. V. S. Ramachandran, *The Tell-Tale Brain: A Neuroscientist's Quest for What Makes Us Human* (New York: W. W. Norton, 2011).

46. Robert Shulman, *Brain Imaging: What It Can and Cannot Tell Us about Consciousness* (New York: Oxford Press, 2013).

47. Thomas West, *In the Mind's Eye* (Buffalo, NY: Prometheus Books, 1991).

48. Doige, *The Brain That Changes Itself*.

49. Klingberg, *The Learning Brain*.

50. Kandel, *In Search of Memory*.

51. David Eagleman, *The Brain: The Story of You* (London: Canongate Books, 2015).

52. "Geometry and Spatial Reasoning," accessed May 6, 2015, http://www.ixl.com/math/geometry-and-spatial-reasoning.

53. Kandel, *In Search of Memory*; Doige, *The Brain That Changes Itself*.

54. Antonio Damasio, *Looking for Spinoza: Joy, Sorrow and the Feeling Brain* (Orlando, FL: Harvest Books, 2003).

55. Damasio, *Looking for Spinoza*.

56. Antonio Damasio and Gil B. Carvalho, "The Nature of Feelings: Evolutionary and Neurobiological Origins," *Nature Reviews Neuroscience* 14 (2013): 143–52.

57. Doige, *The Brain That Changes Itself*.

58. Damasio and Carvalho, "The Nature of Feelings."

59. Lisa D. Wood, "Holding onto Play: Reflecting on Experiences as a Playful K–3 Teacher," *Young Children* (May 2014): 48–56.

60. Narine S. Yegiyan and Andrew P. Yonelinas, "Encoding Details: Positive Emotion Leads to Memory Broadening," *Cognition and Emotion* 25, no. 7 (2011): 1255–62.

61. Rima Faber, "Primary Movers: Kinesthetic Learning for Primary School Children" (master's thesis, American University, 1994).

62. Suzi Tortora, *The Dancing Dialogue: Using the Communicative Power of Movement with Young Children* (Baltimore: Paul H. Brookes Publishing, 2006).

63. Alexander Lowen, MD, *The Betrayal of the Body* (London: Collier MacMillan Ltd., 1967).

64. Federick Perls, MD, *Gestalt Therapy Verbatim* (New York: Batam Books, 1969).

65. Kandel, *In Search of Memory*.

66. Zull, *From Brain to Mind*.

67. Damasio, *Looking for Spinoza*.

68. National Research Council, *How People Learn: Brain, Mind, Experience and School* (Washington, DC: National Academy Press, 2000).

69. John Jonides, "Musical Skill and Cognition," in *Learning, Arts, and the Brain: The Dana Consortium Report on Arts and Cognition*, eds. Carolyn Asbury and Barbara Rich (New York/Washington, DC: Dana Press, 2008), 11–15.

70. "Recall (Memory)," *Wikipedia*, accessed October 21, 2014, http://en.wikipedia.org/wiki/Recall_(memory); Klingberg, *The Learning Brain*.

71. Zull, *From Brain to Mind*.

72. Carter, *Mapping the Mind*.

73. Klingberg, *The Learning Brain*.

74. John Bransford, Linda Darling-Hammond, and Pamela LePage, "Introduction," in *Preparing Teachers for a Changing World: What Teachers Should Learn and Be Able to Do*, eds. Linda Darling-Hammond and John Bransford (San Francisco: John Wiley & Sons, 2005), 1–39.

75. Zull, *From Brain to Mind*.

76. Sarah-Jayne Blakemore and Uta Frith, *The Learning Brain: Lessons for Education* (Malden, MA: Blackwell Publishing, 2005).

77. Darling and Bransford, eds., *Preparing Teachers for a Changing World*, 18–19.

Chapter 7

Imagination and Imagery

Imagination and imagery are symbiotic. It is not possible to create an image without using the imagination, nor is it possible to imagine something without having an image of it. It is not known whether other species have imagination. Perhaps it functions as a survival mechanism. In the evolutionary scale, imagination and imagery are distinctive aspects of human conceptual thought, allowing perception of phenomena not present and conception of possibilities not realized. Symbolic/abstract thought transforms real entities into images and leads to novel conclusions or future predictions.

IMAGINATION

Imagination partners with and is the backbone of creativity. To imagine something is not necessarily a creative act, but creativity requires imagination. Imagination infers stretching beyond what is known to conjure new, unique meanings, linking body, mind, and intellect with the intuition to pursue a vision.[1] The inventor Nikola Tesla credited imagery as the source of his inventions, claiming he could construct and operate devices in his mind to improve them. Beethoven composed symphonies after becoming deaf. He changed music and discarded parts in his mind before writing notes down.[2]

Logic proceeds in a lock-step deductive or inductive progression, whereas imaginative solutions spring to mind.[3] Imagined content can be fantasy, but the convergent thought processes of imagination can also produce beneficial ideas.[4] Imaginative thought passes through stages: framing the problem, a period of exploration and experimentation, and testing of an imagined idea. Sometimes a problem is put aside for incubation, followed by illumination and a solution—phases that can include free association and nonlinear logic.[5]

Brain Network

Creativity increases cerebral activity and blood flow throughout brain regions, representing a highly distributed system. EEG research demonstrates highly creative people, when compared to controls, have greater activity in the right parietal/temporal areas, and higher alpha brain wave activity during periods of inspiration. Alpha activity is dominant in minds open to free thought. Studies on creative problem solving have found lower levels of cortical arousal and stronger alpha synchronization in the central parietal cortices.[6]

Classroom Applications

Young children engage in imaginative play because their brains are not yet hard wired for realism, leaving their minds free to make fanciful associations. Perhaps this is a form of mental scribbling. Children are wonderfully creative until the age of seven or eight, at which time their neural patterns hone in on realism and their drawings begin to look like the real world in form and color.[7]

Studies on arts education identify mental habits connected to the participation in robust arts programs. In addition to learning craft, the arts teach habits of mind and highlight imagination. Arts cultivate mental imagery to guide actions and problem solve.[8] Metaphor is a tool of imagination. It is a creative cognition that relates to established meanings and suggests new ones. Metaphor inspires interpretation rather than using facts to construct learning.[9]

Movement and Dance Connections

It is impossible to create a dance without a synthesis between mind and body. Dance is an embodied realization of imagination. Improvisation is an artistic tool used to explore movement possibilities, experiment with ideas, and search for choreography to communicate artistic intent. A choreographer begins with an inspiration or intent that triggers movement invention. Pioneering modern dancer Hanya Holm instructed her students on the art of dance making by saying dances must be built from within the self, rather than first creating a dance and later deciding what it is.[10]

Movements can relate to real life, but frequently they do not. Famed choreographer Alwin Nikolais explained "The breakdown of story-line, choreographic structure . . . [meant] Time no longer had to support logical realistic events. . . . breaking the barrier of literal time throws the creator into visions . . . motional itineraries way beyond the literal visions."[11] Nikolais also eliminated representational movement as well as literal reference to the human body. His choreography and costumes were visual abstracts of imagined shapes and designs.

Movement Explorations

Imagining the Real: (a) Think about some possible environments: for example, a forest, mountains, lakes, or city. (b) Imagine that you are in each of these different environments. (c) How would you move in each of these environments? (d) How were the movements you created for the different environments the same, and how were they different?

Imagining the Unknown: (a) Imagine that you are in an environment you could not experience in reality: inside a volcano or in outer space. (b) See and feel this imaginary environment. (c) How would you move in it? (d) What caused you to move as you did?

IMAGERY

An image is a mental projection of something realistic or imagined. Imagery has been a subject of inquiry since the Greek philosophers, but until the twentieth century, imagery was connected with introspection. Technology has demonstrated imagery is a component of memory, problem solving, creativity, emotions, and comprehension of language. Mental images created during the initial phase of perception are present, although the stimulus is no longer perceived. These representations preserve properties of the stimulus, giving rise to the subjective nature of perception.[12] Thus, the exact nature of perceptual involvement is open to question.

Brain Network

Daily experiences provide content for imagery. An image is a pattern of electrical energy. The electrical patterns that form an image in our brain are imprinted on the cortex and are different for every experience.[13] An fMRI study showed the left brain encodes categories and creates images based on them, while the right brain generates spatial images and relationships.[14] Sensory experiences are produced by electrical firing within the sensory cortex. The visual sense is pervasive, but patterns of firing are generated by input from all senses, including kinesthetic proprioceptors.[15]

Many people connect mental imagery with vision, but it is possible to imagine sounds, smells, tastes, and body feelings. An fMRI study of adults' auditory and visual imagery found, while sensory imagery activated widely distributed areas of the brain, the vividness of a subject's mental imagery was associated with activity in a modality-specific brain network. During the auditory portion, subjects recalled familiar melodies, and in the visual section they focused on imaging common objects. Auditory images deactivated the visual neural network with a parallel deactivation of the auditory network during visual imaging.[16]

Classroom Applications

Imagery is a powerful learning tool, making content concrete and easier to remember. Remembering vocabulary is facilitated if words are visualized as objects or are related to images.[17] The recall mechanism of mnemonics is strengthened by using associated images. For example, imagine a house or castle and associate data with rooms or décor. Assigning colors to content is another technique, as is using humorous images.

Movement and Dance Connections

Imagery inspires improvisation and intensifies performance, technique, and movement dynamics. It kick starts the brain into new patterns. Imagery can introduce qualities of movement into the lexicon, and provide new experiences to widen movement vocabulary.

Early in the twentieth century, Florence Fleming Noyes developed a system of teaching dance that focused on imagery from nature to free women of their physical inhibitions. Images such as the wind, rustle of leaves, or clouds helped her students discover creative freedom and fluidity.[18] Dance teachers use images to stimulate alignment and correct technique.[19] Graham brought a high level of concentration to students, company members, and her roles using imagery to release inner energy.[20]

Movement Explorations

Visual Image: (a) Select a painting and carefully study it until you can visualize all of its parts and any characters within it. (b) Imagine that you are within the painting. (c) Create movements as you thoroughly explore being in the painting.

Kinesthetic Image: (a) Pretend that you are walking on silk and stroking the silk with each footstep. (b) As you walk forward, continue to focus on this image. (c) How did your feet feel as they came in contact with the floor? Did the feelings change how they moved?

Auditory Image: (a) Imagine a sound, and move your arm or body in a way that responds to this sound. (b) Imagine a second sound opposite in quality to the first and respond to this sound. (c) Did the second movement you created differ from the first? If so, how did it differ?

Images and Moving Efficiently: (a) Walk forward while thinking of your body as the bow of a ship cutting through water. (b) How did this image change the quality of your walk? (c) How did it feel in your body to walk in this way?

VISUAL IMAGERY

The content of visual imagery is apparent by picturing something in your mind—its color, relationships of line, design, shape, or location, and the visual aspects of texture. Visual imagery is like a series of mental photographs and is spatial in nature.[21] Images can be generated, inspected, maintained, and transformed.[22]

Gardner's concept of spatial intelligence focuses on visual imagery as the capacity to accurately perceive the visual world; make transformations and modifications of perceptions; and recreate aspects of visual perceptions in absence of original stimuli.[23] Gardner noted Picasso as an example of visual genius because he was skilled in noticing details and could think in spatial configurations.[24] Temple Grandin, an autistic professor, claims the pictorial nature of her thinking enabled her to design humane equipment used with animals, especially cattle.[25]

Visual imagery reproduces a model of the environment, making it possible to mentally rehearse future actions to perfect a desired performance. For instance, sports team members mentally review a plan of action immediately before a play.

Brain Network

Visual, spatial imagery is created from input stored in long-term memory that corresponds to a mental map of objects and their locations. Stimuli activate the occipital region in the brain and a pathway that runs from the occipital lobe to the posterior parietal lobe. This portion of the parietal lobe processes visual properties such as location, size, and orientation.

Classroom Applications

Teachers rely on visual imagery. Young children are taught to visualize words from picture books, but as they mature illustrations lessen and students learn to create visualizations in their own minds from the written word. Visual learners often have difficulty learning auditory language skills. The Viconic Language Method has proven effective; it superimposes visual pictures onto auditory language, translates English into visual thinking, and uses cartoon text bubbles to depict language in drawings.[26] The proliferation of visual imagery in our increasingly screened world has encouraged some educators to explore visual imagery in their classrooms.

Movement and Dance Connections

Since its prehistoric roots, tribal dance has embodied images of animals or gods, both fierce and benevolent. Ballet, with its wing-shaped arms, high leg extensions, and ariel leaps, projects images of flight. In modern dance, Alvin Ailey's dance *Revelations* begins with dancers reaching upward, and using imagery of angels soaring toward heaven. The famous section of this dance, "Waves in the Water," depicts bodies liquidly undulating as an embodiment of waves.

Visual imagery is used in teaching dance skills. Imagining a bouncing ball or deer bounding over a hedge heightens buoyancy in jumps. The image of a plumb line extending down through the body into the earth centers rotational movements.[27]

British choreographer Wayne McGregor encouraged his dancers to creatively problem solve using visual imagery. His visual images included spatial images and interacting with manipulating three-dimensional objects. Images were colored by the emotional tone of a narrative. This study showed while the dancers thought they were focusing on the physical aspects of their movements, they were actually focusing on its conceptual aspects.[28]

Movement Explorations

Spatial Image: (a) Close your eyes and imagine you are walking through your home. (b) Create a mental map of the path you just walked. (c) In a large, clear space, move along the pathways and directions of the mental map you made.

Imagined Relationships: (a) Imagine a relationship that can exist between your body and a chair (e.g., next to, over or under the chair). (b) Recreate this physical relationship with the chair. (c) Imagine a second relationship and recreate it. (d) Did the physical relationships feel different from their images? Did the two different relationships feel differently?

Manipulation of Visual Image: (a) Reimagine the chair from exploration number two. (b) Begin by thinking of the chair as right side up and make the shape of the chair with your body. (c) In your mind, manipulate the position of the chair so that it is no longer upright; for instance, lying on its side or facing down to the floor. (d) Change your body orientation to reflect the change of the chair. (e) Does your body feel differently with the change?

KINESTHETIC IMAGERY

Kinesthetic imagery or the bodily feeling of a movement is constructed from feedback generated in proprioceptors located in joints, tendons, and muscles.

Rather than a focus on the external environment, the image is constructed from proprioceptive feedback integrated in the brain.[29]

Kinesthetic images produce bodily sensations that accompany performance of an action. To move lightly like a floating feather is a kinesthetic image. In a visual image, the focus is on a mental picture of a feather floating, while the focus for the kinesthetic image is on body sensations. Kinesthetic imagery created the internal dynamic stillness shown in figure 7.1.

Figure 7.1 Kinesthetic imagery in action. *Source:* Reprinted by permission, from S.C. Minton, 2007, *Choreography: A Basic Approach Using Improvisation*, 3rd ed. (Champaign, IL: Human Kinetics). Photographer Joe Clithero of B & J Creative Photography.

The kinesthetic sense is probably the least understood of all the sensory systems. American society values mental processes most highly in education. Society is goal versus process oriented. People are concerned with how they look externally rather than valuing internal awareness. Proprioceptive awareness is sometimes described as the forgotten sense, and it is not always included.[30]

Brain Network

Kinesthetic receptors include different structures: muscle spindles (fluid-filled capsules within muscles); golgi tendon organs in tendons near muscles; several types of joint receptors inside joints; and skin receptors[31] (see figures 7.2–7.4). Muscle spindles along with intrafusal fibers respond to quick and maintained stretching and are sensitive to changes in velocity. Golgi tendon

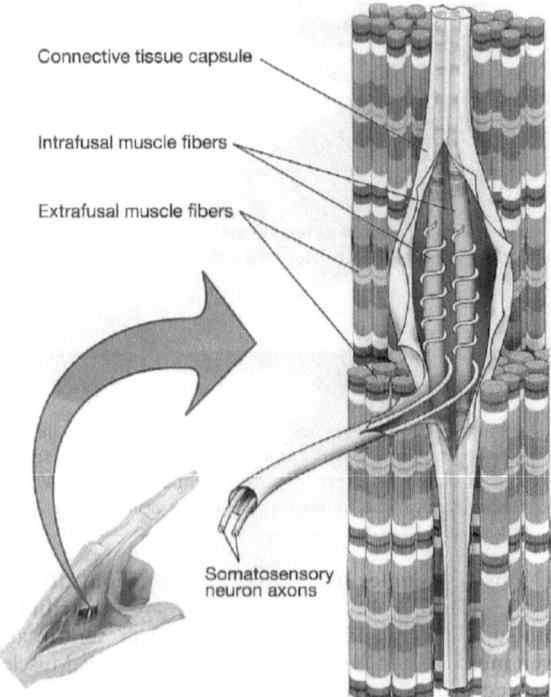

Figure 7.2 Muscle spindle. *Source*: Schenkman, Margaret; Bowman, James; Gisbert, Robyn; Butler, Russell, *Clinical Neuroscience for Rehabilitation*, 1st © 2013. Reproduced by permission of Pearson Education, Inc., New York, New York.

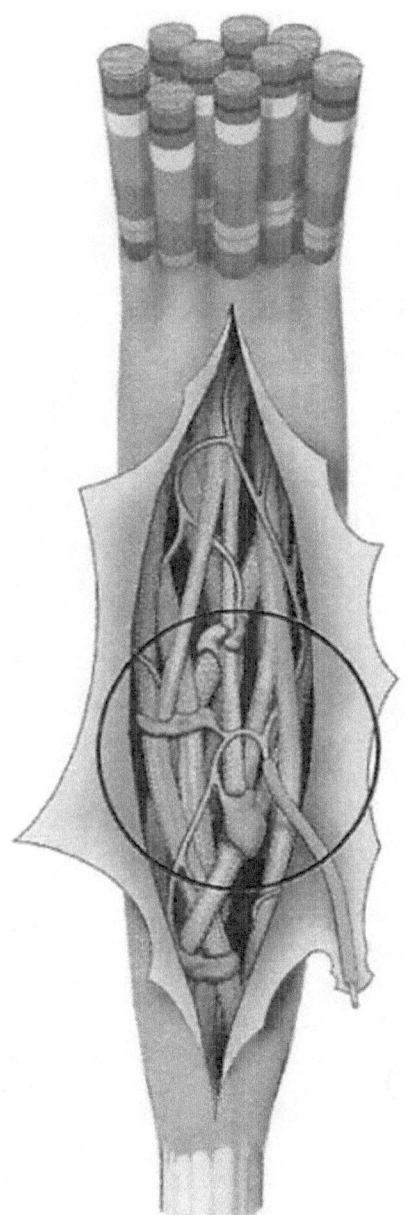

Figure 7.3 Golgi tendon organ. *Source*: Schenkman, Margaret; Bowman, James; Gisbert, Robyn; Butler, Russell, *Clinical Neuroscience for Rehabilitation*, 1st © 2013. Reproduced by permission of Pearson Education, Inc., New York, New York.

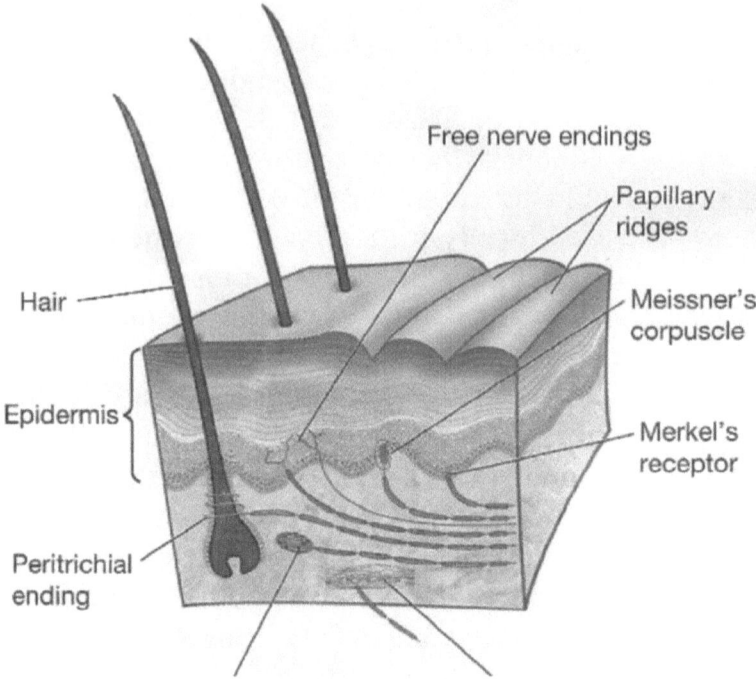

Figure 7.4 Skin receptors. *Source*: Schenkman, Margaret; Bowman, James; Gisbert, Robyn; Butler, Russell, *Clinical Neuroscience for Rehabilitation*, 1st © 2013. Reproduced by permission of Pearson Education, Inc., New York, New York.

organs sense tension or active muscular contraction and velocity of tension changes, but they are less responsive to passive stretching. Joint receptors respond to joint position and tension changes. Skin receptors sense vibration, displacement, and tissue damage.[32]

The muscle spindles and golgi tendon organs are connected to the cerebellum, but the spindles may also be part of a feedback loop between the motor cortex and muscles.[33] Information from the golgi tendon organs may be channeled to the cortex.[34] All of these structures together with the vestibular apparatus are responsible for producing kinesthetic sensations of body position and movement.

The sensory and motor cortices mentioned above lie at the juncture between the frontal and parietal lobes.[35] The motor cortex processes motor information, while the somatosensory cortex processes tactile information. Cutaneous input is projected from the thalamus to the somatosensory cortex, preserving topographic organization in the somatic cortex.[36] The sensory and motor cortices include divisions that represent parts of the body, so specific brain areas control specific body parts. This representation of body parts in the brain, the cortical homunculus, is mapped on the postcentral gyrus[37] (see figure 7.5).

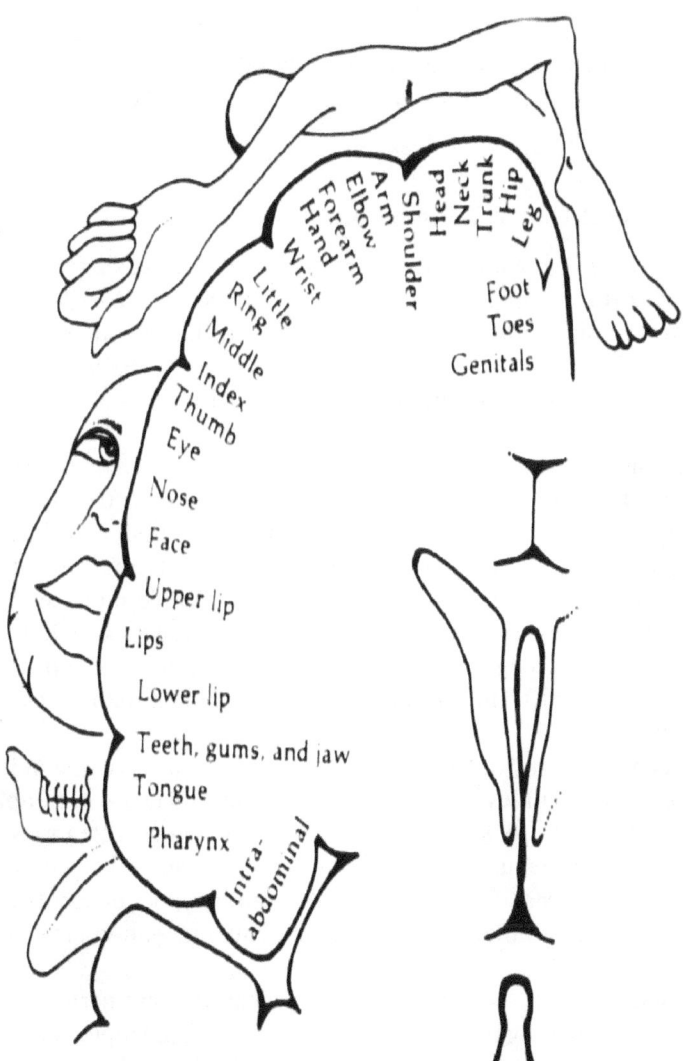

Figure 7.5 Cortical homunculus. *Source*: From *The Tell-Tale Brain: A Neuroscientist's Quest for What Makes Us Human* by V. S. Ramachandran. Copyright © 2011 by V. S. Ramachandran. Used by permission of W. W. Norton & Company, Inc.

Classroom Applications

An integration of kinesthetic imagery in the classroom unites the body and mind for holistic understanding of content. Children can learn to write letters by making their shapes with their bodies and then tracing them with fingers. Addition can be understood by walking forward on a number line, and subtraction by traveling backward to spatially experience numbers getting greater or smaller.

Blind students are dependent on kinesthetic/tactile and spatial imagery. In the absence of visual information, they build mental maps of routes based on visceral memories. Reading Brail depends on kinesthetic, tactile, and spatial maps from fingers.

Movement and Dance Connections

Kinesthetic images breathe life into movement. In class or rehearsal, when dancers perform movements repeatedly, it can become a drill to approach practice from only a physical perspective. Use of kinesthetic imagery keeps movement alive as a recreation through inner body feelings. Kinesthetic images are also reflected in the body and brain of observers watching dance due to the empathetic response of mirror neurons.

Studies designed to compare the effectiveness of visual versus kinesthetic imagery show kinesthetic imagery is more effective in perfecting the timing or duration of the target movements, while visual imagery is more helpful when learning movement skills that involve form reproduction.[38] Another study videotaped college jazz and modern dance teachers to determine their imagery preferences. Intermediate classes included greater use of imagery than beginners, especially visual imagery in comparison to kinesthetic imagery.[39] This could reflect the dominant nature of the human visual system.

Researchers in England conducted a study to analyze the relationship between dancers' skill level and their ability to use imagery. A sample of 250 dancers ages sixteen to sixty-six was sorted into three categories: leisure-time dancers, intermediate-level training full time, and elite preprofessional or professional dancers. It was found that the more experienced dancers used images of greater complexity and became deliberate and controlled in their imagery use.[40]

Kinesthetic imagery is effective for both learning and performing dance. To capture body feelings instead of depending on visual input, some teachers turn their classes away from mirrors traditionally located in the front of the studio. Of course, there are no mirrors during performance, so dancers rely largely on their kinesthetic sensations.

Movement Explorations

Kinesthetic Image: (a) Picture clouds as you would see them from an airplane. (b) Imagine you are floating through the clouds. Then, move as if you were walking or swimming through the clouds. (c) Did these images affect the way you moved?

Proprioceptive Image: (a) Tighten the muscles of your arm and focus on the body feeling this action creates. (b) Create a movement that captures this same body feeling. (c) What type of movement would you create that captures a relaxed body feeling?

Recapturing Kinesthetic Imagery: (a) Perform a different movement that has the same tightened body feeling as the first movement you created in the previous exploration. (b) Create a movement with a different body part that has the same kinesthetic feeling. (c) What was your point of focus that enabled you to recreate the same body feelings found in the two movements?

Audience Response: a) Watch a short YouTube video of a modern dance performance that is slow and smooth. (b) What is your body response to viewing this dance? (c) Create movements that duplicate this second body feeling.

BODY IMAGE

Body images or schema are mental representations that shape one's physical identity. It governs experience of the body; how people think they look, alignment, tension, breath, and movement patterns.[41] This is constructed over time from experiences as early as infancy.

Body identity is created through social filters from family, communities, ethnic groups, or religions. A positive body image begins in infancy with positive feedback. As a person develops, their body image may not match their physical body. If a person develops a sense their body is inferior, the image becomes negatively distorted.[42]

In America, the ideal female body is projected by media as extremely thin with large breasts, while the male ideal is lean and muscular. The first focuses on sexual appearance and how a woman looks, in contrast to the second, which emphasizes power and how a man's body functions. Based on these differences, women are more likely to diet and have cosmetic surgery, while men use protein supplements, steroids, and weight-bearing exercises.[43]

Appearance and body satisfaction are major contributors to overall self-esteem since a sense of one's body is a central organizing component involving interplay between cognitive, emotional, and behavioral processes.[44]

A negative image can reduce confidence, inhibit physical achievement, and even interfere with social interaction.

When part of the body is damaged or amputated, there are mental repercussions. Rehabilitation must address these issues as well as physical function.[45] Techniques have been developed using mirrors and other technology to trick the brain into recreating a complete and whole body image.[46]

Brain Network

Studies using fMRI identified brain regions that affect body image. The stimuli varied, but were, in general, those that could elicit a poor body image such as pictures showing exaggerated body parts or negative words. Subjects' fusiform gyrus, part of the brain involved in face and body recognition, was activated, along with possible increased activation of the amygdala and anterior cingulate cortex, areas associated with experiencing emotion and fear.[47]

Classroom Applications

Schools are a powerful environment in which a positive body image can be promoted since a student's self-concept affects their performance and social interaction in the classroom. Students who are self-conscious are less likely to participate in and benefit from classroom activities. Children develop their body image in relation to how their bodies function. Noncompetitive exercise such as dance can improve self-esteem even if it does not produce significant physical changes.[48]

While promoting an acceptance of diversity, schools can help students develop an appreciation for all body types, build awareness of biases fostered by media, and reduce the value placed on competitive physical achievement. Schools can also educate students about nutrition and healthy practices, rather than dieting or focusing on their looks.

Movement and Dance Connections

Dance is an artistic form that provides exercise and functional understanding of one's body. When approached with an open perspective, dance class fosters a positive body image and appreciation of every body type. Unfortunately, an aesthetic partial to long-limbed, skinny dancers with a long neck, small bust, and articulate feet has become paramount for a career dancer or teacher.

Disturbance of one's body image can produce psychological and physical problems, such as anorexia and bulimia, particularly in adolescent girls who are overly concerned with shape or weight. A symptom is an intense fear of gaining weight, even when underweight. Anorexics starve themselves. Bulimics binge eat and then purge by vomiting or misusing laxatives, diuretics, or enemas to rid the body of what has been eaten.[49]

A Canadian study of ballet students reported approximately 6.5 percent suffered from anorexia—a percentage of occurrence higher than women in the general population of similar age.[50] Another study examined the effect of mirrors on the body image of female college students in beginning ballet. One group was taught using mirrors, while the other group was taught by the same instructor, but with no mirrors. As a result, the students who danced without mirrors had greater gains in body area satisfaction and movement skills.[51]

As dance educators and choreographers have become increasingly aware of the dangers of body image disturbances, a growing number of studios and companies are appreciating the healthy looking dancer. In the general population, obesity is becoming a cultural health problem. Stress on healthy eating habits rather than weight is becoming a desired value. Dancers rely on a healthy body as an efficient tool as well as an aesthetic instrument. The dancer becomes the dance through their body.

Movement Explorations

Body Satisfaction: (a) Sit in a quiet location and do a mental scan of your entire body part by part. (b) Which parts of your body do you view in a satisfied way? (c) Which parts of your body are less satisfying? (d) Why do you suppose you feel as you do about the different parts of your body?

Social Norms: (a) Select a popular magazine and look through its pages to determine what type of body is presented in most of the photos. (b) What conclusion concerning body type did you come to? (c) Do you think the images presented in the magazine affect your body image and the body image of others? (d) Provide a reason for your answer to this question.

NOTES

1. Elliot W. Eisner, *The Arts and the Creation of Mind* (New Haven, CT: Yale University Press, 2002); Robert and Michèle Root-Bernstein, *Sparks of Genius: The 13 Thinking Tools of the World's Most Creative People* (Boston: Houghton Mifflin, 1999), 5.

2. Root-Bernsteins, *Sparks of Genius*.

3. Michael Michalko, *Cracking Creativity: The Secrets of Creative Genius* (Berkeley, CA: Ten Speed Press, 2001).

4. Mihaly Csikszentmihalyi, *Creativity: Flow and the Psychology of Discovery and Invention* (New York: Harper Perennial, 1996).

5. Richard L. Gregory, ed., *The Oxford Companion to the Mind*, 2nd ed. (New York: Oxford University Press, 2004), 442–47.

6. Richard J. Haier and Rex E. Jung, "Brain Imaging Studies of Intelligence and Creativity: What Is the Picture for Education?" (lecture, Creative Brain Conference, Washington, DC, May 7–9, 2009).

7. Howard Gardner, *Artful Scribbles* (New York: Harper Collins, 1980).

8. Ellen Winner and Lois Hetland, "Art for Our Sake: School Arts Classes Matter More Than Ever—but Not for the Reasons You Think," *Boston Globe*, accessed December 9, 2014, http://www.boston.com/news/globe/ideas/articles/2007/09/02/art_for_our_sake/.

9. Arthur D. Efland, "The Arts and Six Revolutions in Cognition: Implications for Arts Education and Schools" (lecture, Creative Brain Conference, Washington, DC, May 7–9, 2009); Jay Seitz, "The Development of Metaphoric Understanding: Implications for a Theory of Creativity," *Creativity Research Journal* 37, no. 10 (1997): 347–53.

10. Jean M. Brown, Naomi Mindlin, and Charles H. Woodford, eds., *The Vision of Modern Dance*, 2nd ed. (Hightstown, NJ: Princeton Book Company, 1998), 77.

11. Brown, Mindlin, and Woodford, *The Vision of Modern Dance*, 117.

12. Stephen Kosslyn, William L. Thompson, and Giorgio Ganis, *The Case for Mental Imagery* (New York: Oxford University Press, 2006).

13. James E. Zull, *From Brain to Mind: Using Neuroscience to Guide Education* (Sterling, VA: Stylus Publishing, 2011).

14. Stephen Kosslyn, accessed June 11, 2015, http://en.wikipedia.org/wiki/Stephen_Kosslyn.

15. Zull, *From Brain to Mind*.

16. Mikhail Zvyagintsev, Benjamin Clemens, Natalya Chechko, Krystyna A. Mathiak, Alexander T. Sack, and Klaus Mathiak, "Brain Networks Underlying Mental Imagery of Auditory and Visual Information," *European Journal of Neuroscience* 37 (2013): 1421–34.

17. Sarah-Jayne Blakemore and Uta Frith, *The Learning Brain: Lessons for Education* (Malden, MA: Blackwell Publishing, 2005).

18. Rima Faber, "Primary Movers: Kinesthetic Learning for Primary Age Children" (master's thesis, American University, 1994).

19. Eric Franklin, *Dance Imagery for Technique and Performance* (Champaign, IL: Human Kinetics, 1996).

20. Marian Horosko, *Martha Graham: The Evolution of Her Dance Theory and Training, 1926–1991* (Pennington, NJ: A Capella Books, 1991).

21. Oliver Sacks, *The Mind's Eye* (New York: Vintage Books, 2010).

22. Kosslyn, Thompson, and Ganis, *The Case for Mental Imagery*; V. S. Ramachandran, *The Tell-Tale Brain: A Neurologist's Quest for What Makes Us Human* (New York: W. W. Norton, 2011).

23. Howard Gardner, *Frames of Mind: The Theory of Multiple Intelligences* (New York: Basic Books, 1983).

24. Howard Gardner, *Creating Minds* (New York: Harper Collins, 1993).

25. Temple Grandin, *Thinking in Pictures: My Life with Autism* (New York: Vintage Books, 2006).

26. Ellyn Arwood and Carole Kaulitz, "Language Strategies for Learning with a Visual Brain" (lecture, Creative Brain Conference, Washington, DC, May 7–9, 2009).

27. Franklin, *Dance Imagery*.

28. Glenna Batson with Margaret Wilson, *Body and Mind in Motion: Dance and Neuroscience in Conversation* (Bristol, UK/Chicago: intellect, 2014).

29. Zull, *From Brain to Mind*.

30. Sally S. Fitt, *Dance Kinesiology*, 2nd ed. (New York: Schirmer, 1996).

31. Sandra Minton and Jeffrey Steffen, "The Development of a Spatial Kinesthetic Awareness Measuring Instrument for Use with Beginning Dance Students," *Dance: Current and Selected Research* 3 (1992): 73–80.

32. Glenna Batson, "Update on Proprioception Considerations for Dance Education," *Journal of Dance Medicine & Science* 13, no. 2 (2009): 35–41.

33. George H. Sage, *Introduction to Motor Behavior: A Neuropsychological Approach*, 2nd ed. (Reading, MA: Addison-Wesley, 1977).

34. Marjorie Moore, "Golgi Tendon Organs: Neuroscience Update with Relevance to Stretching and Proprioception in Dancers," *Journal of Dance Medicine & Science* 11, no. 3 (2007): 85–92.

35. Rita Carter, *Mapping the Mind* (Berkeley, CA: University of California Press, 2010).

36. Sage, *Introduction to Motor Behavior*.

37. Ramachandran, *The Tell-Tale Brain*.

38. Khaled Taktek, Nathaniel Zinsser, and Bob St-John, "Visual Versus Kinesthetic Mental Imagery: Efficacy for the Retention and Transfer of a Closed Motor Skill in Young Children," *Canadian Journal of Experimental Psychology* 62, no. 3 (2008): 174–87.

39. Sandra Minton, "Assessment of the Use of Imagery in the Dance Classroom," *Impulse* 4 (1996): 276–92.

40. Sanna M. Nordin and Jennifer Cumming, "The Development of Imagery in Dance Part II: Quantitative Findings from a Mixed Sample of Dancers," *Journal of Dance Medicine & Science* 10, nos. 1 & 2 (2006): 28–34.

41. Franklin, *Dance Imagery*.

42. Mirka Knaster, *Discovering the Body's Wisdom* (New York: Bantam Books, 1996).

43. Sarah K. Murnen, "Gender and Body Images," in *Body Image: A Handbook of Science, Practice, and Prevention*, 2nd ed., eds. Thomas F. Cash and Linda Smolak (New York: Guilford Press, 2012), 263–70.

44. Marika Tiggemann, "Sociocultural Perspectives on Human Appearance and Body Image," in *Body Image*, 12–19; Thomas Cash, "Cognitive-Behavioral Perspectives on Body Image," 39–47 in *Body Image*.

45. Oliver Sacks, *A Leg to Stand on* (New York: Simon Schuster, 1984).

46. Norman Doidge, *The Brain That Changes Itself*.

47. Jessica L. Suisman and Kelly L. Klump, "Genetic and Neuroscientific Perspectives on Body Image," in *Body Image*, 29–38.

48. Kathleen A. Martin Ginis and Rebecca L. Bassett, "Exercise and Changes in Body Image," in *Body Image*, 378–86.

49. Eleanor H. Wertheim and Susan J. Paxton, "Body Image Development in Adolescent Girls," 76–84; Sherrie Selwyn Delinsky, "Body Image and Anorexia Nervosa," 279–87; Janis H. Crowther and Nicole M. Williams, "Body Image and Bulimia Nervosa," 288–95 in *Body Image*.

50. Bonnie E. Robson, "Disordered Eating in High School Dance Students: Some Practical Considerations," *Journal of Dance Medicine & Science* 6, no. 1 (2002): 7–13.

51. Sally A. Radell, Daniel D. Adame, and Steven R. Cole, "The Impact of Mirrors on Body Image and Classroom Performance in Female College Ballet Dancers," *Journal of Dance Medicine & Science* 8, no. 2 (2004): 47–52.

Chapter 8

Learning

The College Board reported students taking dance for four or more years averaged math and verbal scores twenty-seven points higher than students not involved in its study. Seattle third-graders who study language arts using dance boosted their scores on the Metropolitan Achievement Test reading scores by 13 percent. Dance develops sequencing, collaboration, interpretation, coherence, and problem-solving skills.[1] In addition, dancers learn to work hard and strive toward excellence. They acquire thinking skills, social skills, and a work ethic for learning academics.

ACTIVE LEARNING

At the advent of the twentieth century, John Dewey ushered in a progressive philosophy of education by advocating learning through educational experiences. Dewey was pivotal in shaping the idea that learning is most effective through active involvement as opposed to passively sitting. His vision of education related to children's health and welfare outside the classroom. Dewey professed school should reflect life, rather than prepare for life; that factory-model education disconnects from outside experiences, and fails to apply to learning.[2]

Dewey's methods could not accommodate the flood of immigrant children entering American public schools in the early 1900s. The factory model of rote education therefore predominated for the rest of the century. Understandings of cognitive science and brain research in the new millennium support active learning and renewed interest in Dewey's philosophy.

Brain Network

Active learning is a whole-brain phenomenon. It simultaneously links and reinforces experiences in multiple brain areas: concrete sensory information in the parietal lobe, abstract concepts in Broca's and Wernicke's left temporal areas, spatial relationships in the right temporal lobe, and emotions in the amygdala. Their orchestration stimulates the hippocampus, which improves memory.

Classroom Applications

There is a common saying attributed to Confucius, "I hear and I forget. I see and I remember. I do and I understand."[3] Active learning deepens learning experiences. Memorizing content or observing information from afar does not internalize content with meaning, and it is not a route toward understanding.[4]

Active constructive teaching strategies lead to deeper understanding[5] because it promotes understanding, problem-solving skills, stimulates curiosity and independence, and creates a positive learning environment.[6] David Perkins developed Teaching for Understanding that addresses the "idea-action gap." This framework emphasizes active learning as students discover knowledge and apply it to extend meaning.[7] Geometry involves using and understanding the function of shapes, not just recognizing them (e.g., wheels on a car are a circle so they roll smoothly, a house is a rectangle to make efficient use of interior spaces).

Students need opportunities to learn from trial and error because it forms connections between stimuli and neural responses. Multiple successful trials that achieve a desired result are reinforced by stamping out unsuccessful trials and imprinting successful attempts into the brain.[8]

Movement and Dance Connections

Dance embodies learning through action and can be used to teach other academic disciplines. Ideas expressed in movement impart meaning to content. *Science with Dance in Mind*, a project of the Primary Movers with Baltimore County, paired five K–3 classrooms with a dance specialist to teach science curriculum through dance movement experiences. Assessment demonstrated that most students retained greater scientific knowledge and widened their understanding of movement.[9]

An interdisciplinary project was conducted at a performing arts charter school with ninth- and tenth-graders to bring dance and anthropology together. In the project, hip hop dance was used to explore localization and homogenization. The students were expected to determine if hip hop takes on attributes of a local culture when it is performed outside America, or if a cultural group absorbs hip hop as it exists in the United States. After learning some hip hop moves and pertinent anthropological concepts, the students concluded hip hop has a different purpose when performed by various cultural groups.[10]

Scientists conducted an experiment demonstrating that movement is an effective teaching strategy, especially for physically active students. Professional dance company members explored a living cell. The most abundant molecule in a cell is water, which moves at high speed. This produces collisions and gives rise to diffusion or a random movement of molecules. The dancers demonstrated collision and diffusion, and showed movement can be a useful complement to computer simulation when testing hypotheses.[11]

In a similar vein, noted choreographer Liz Lerman partnered with scientists to choreograph a dance, *Ferocious Beauty: Genome*, based on the genetic DNA chain. She successfully worked with four thousand students at James Madison University.[12]

Movement Explorations

Curiosity: (a) Research a culture that interests you. Articles, books, print media, and the Internet can be used in your research. (b) Write down details you find about the concept. By researching the culture in this way, you are constructing your understanding of it. (c) Create a simple movement based on selected details. (d) Did creating the movements embody your understanding of the culture? Did using movement provide a different perspective?

Active Learning: (a) Select a specific historical period. (b) Read about the ideas, trends, and events that motivated thoughts and actions during the time period. (c) Select one idea that you think represents the era as it is described in the literature. (d) Create a simple action that you feel captures the feeling or spirit of the idea. (e) Explain why you connected the movement you created to your selected idea.

Improve Understanding: (a) Find a dance on YouTube based on factual information or an historical event. (b) What was the perspective communicated in the dance? (c) Did this perspective add to your understanding?

EMBODIED COGNITION

There is a branch of cognitive science called embodied cognition that recognizes thought is unified with the body; that it does not arise from the brain or the body alone, but is shaped by experiences enacting with the body.[13]

Learning depends on developing neural connections in the brain that require learners to engage with the world through action.[14] Embodied actions adapt to needs, capture feelings, or integrate the spirit of something outside the body of the mover. Embodied learning relies on connecting contextual information and the body's kinesthetic sense.

Brain Network

The structure of the human brain supports the connection between mind and body. The cerebellum, which coordinates movement, connects with the prefrontal cortex by a trunk of nerve cells. Movement activates the brain in a full array of cognitive functions.[15] Mirror neurons play a role in embodied learning, which includes the ventral and dorsal premotor cortex and several regions in the parietal cortex.[16]

Classroom Applications

Through embodied learning, students actively construct knowledge, build their own understanding of content, and creatively connect information with experiences.[17] Embodied exploration leads to multiple insights, improved understanding, and the ability to communicate understanding to others.[18] It is a natural teaching strategy supported by neurological research, yet it is not included in most curricula.

Two Canadian professors described how teaching a second language was made more effective using innate bodily learning tools. It proved successful because it incorporated the senses, emotions, patterns, and humor that are primal and central learning mechanisms. Humor tickles the brain and captures attention with novelty, flexibility of mind, and imagination.[19]

Movement and Dance Connections

Dewey's influence inspired educators and dancers to integrate dance into education as a powerful ally. The current practice of teaching academics through the arts highlights the effectiveness of dance integration. This holds true if high standards are met in all disciplines so students learn about the art of dance as well as required curricula in other subjects. Dance is far more encompassing than acting out an idea without words, a simplistic technique too often cited as dance integration.

The "Basic Reading through Dance" program, executed in Chicago Public Schools with hundreds of children, was highly successful. It taught letter recognition, phonemes, spelling, and reading comprehension. Quantitative assessment and analysis found greater improvement in children using movement as their learning tool than for a control group using no movement.[20] Research with secondary school populations has attested that the majority of children from African American roots and Asian and Hispanic minority backgrounds are kinesthetic learners.[21]

Embodied writing is a technique used to improve students' writing abilities and develop awareness of dance. Students experience an external phenomenon, internalize it through body movement, and recreate their experiences through writing. Their papers showed improvement in descriptive and analytic writing skills.[22]

Movement Explorations

Contextual Learning: (a) Select a word or concept that could apply to or be used in a number of different contexts or situations. For example, the word *leaf* can refer to a part of a plant, a sheet of paper that serves as a page in a document or book, a thin sheet of metal, or a hinged, flat part of a door or table. (b) Analyze the different meanings of the word or concept you have selected. (c) Use a movement or a series of movements to communicate the different meanings of your concept or word.

Embodied Understanding: (a) Select a poem. (b) Read the poem and interpret it through movement several times. (c) Did dancing the poem help you remember the poem? Did you experience new or multiple insights from dancing the poem?

Transfer to Embodied Writing: (a) View a short YouTube clip of a dance. (b) Write a description of some movements in the dance. (c) Have a partner read what you have written, and create movements from your description. (d) Do the movements your partner created look like the movements in the original YouTube dance? How are they similar? In what ways are they different?

PATTERNS

Patterns are the result of combining simple elements into expressive relationships between parts.[23] Lines and angles have a relationship with one another whether or not the relationship is repetitive within the design, but the term *pattern* commonly refers to a discernable regularity.

Rhythmic patterns are a repetition of auditory components. Wallpaper or a textile print will often have a repetitive visual pattern. Chaotic patterns involve fractals, a geometric structure that has an uneven or irregular shape repeated in progressive sizes. In psychology, a pattern refers to the repetition of behaviors. In dance and visual arts, the term *pattern* is used in two ways, as a single or repetitive design.

Brain Network

The brain loves patterns and is wired instinctively to recognize and form them. It gathers input from the environment and creates patterned brain networks used to predict accurate responses to new input. New patterns are connected to existing neuron pathways for the purpose of comparing them to stored patterns and past experiences.[24] The electrical patterns fired in the brain are vehicles of meaning, so new patterns carry new meaning.[25] The flow of patterns and patterned combinations supports comprehension and creativity.[26]

During learning, brain dendrites grow and form connections at their synapses with other neurons, creating neural networks.[27] Branching neurons form physical patterns from one neuron to the next. Chunking information links patterns and aids memory.

In problem-solving, patterns of electrical activity in the front integrative cortex are selected as images. They form as short-term memory, and travel to link with other neurons to produce new images. Eventually, the new images created become physically wired. There is a limited amount of information that can be held in short-term memory, so chunking neural patterning increases capacity.[28]

Classroom Applications

Effective learning and teaching depends on strategies that encourage new neural patterns.[29] Decoding information, problem solving, and learning requires recognizing, forming, and extending existing patterns.

Music offers an enjoyable way to anchor information through patterns of melody and rhythm. Singing directions or academic content is effective for students with learning disabilities. Preschoolers learn the alphabet song. Oliver Sacks recounts an anecdote about a student who quoted his lectures and other resources on an exam by setting the texts to song.[30]

Movement and Dance Connections

Movement patterns are the basis of dance. Repetitive rhythms in tribal dance can transport a person into a trance. Patterns in time are created by moving on the beat, faster, slower, or in contrast to a steady pulse beat, creating both even and syncopated (uneven) patterns. Rhythms in percussive dance are audible.

Rhythmic patterns can form communicative images. Military drumming is a strong, driving beat that rallies soldiers and intimidates enemies. African polyphonic rhythms establish community between musicians and dancers. Hip hop inherited complex rhythms from its African roots. Spatial and visual patterns are created from levels, body shapes, locations, or pathways as dancers move.

A dancer's changing relationships to objects or other dancers form dramatic, entertaining, or comic patterns. Dynamic patterns are formed by changing intensity or energy quality that add interest and communicate meaning. Different dance genres have defined dynamic characteristics and patterns. Hip hop and jazz are high energy, fast paced and strongly rhythmic. Square dance is based on geometric patterns in space. The choreographic devices mentioned previously are built from repetitive, relational patterns.

Using positron-emission tomography (PET), researchers found repetition of a spatial pattern, the dance box step, activated part of the parietal lobe responsible for spatial perception, orientation, and body control. The area activated is close to where kinesthetic representations reside. The researchers concluded activation of this part of the parietal lobe serves to plot a spatial path from an egocentric perspective, and entrains muscles to perform movement patterns.[31]

Movement Explorations

Rhythmic Patterns: (use head, shoulders, or a variety of body parts): (a) Find a comfortable, steady beat. Move faster than it, slower than it, or syncopate it by adding to or leaving out some of the beats. (b) Create several patterns using this method. (c) Draw a diagram of the rhythmic patterns you created. (d) Perform each of these patterns using a different body part.

Spatial Patterns: (a) Explore a variety of movements based on levels, shapes, locations, or pathways. (b) Create a spatial design or pattern using one of these components. (c) Create a second spatial pattern using another component. (d) Compare your ability to perform each of the spatial patterns.

Relational Patterns: (a) Either work with another person or choose an object. Move in relation to your chosen subject. For example, you can move over, beside, in front, or behind the object or person. (b) Create a relational pattern by changing your relationship to the stationary person or object. (c) Perform the relational pattern you created.

Dynamic Patterns: (a) Develop a sequence or phrase of movement. (b) Create a dynamic pattern by changing your use of intensity and quality. (c) Perform the dynamic pattern you created.

ASSOCIATIONS

The brain encourages formation of associations to facilitate learning by bonding new and familiar information into memory. An association is formed when information is recognized and linked to previously stored memories. Although the mind functions as a network system, it builds knowledge in a progressive process.

Brain Network

The brain's integrative cortex creates associations dispersed over the back and front of the brain. The right hemisphere provides the big picture, and the left hemisphere manipulates details. The integrative cortex is differentiated and larger than the sensory or motor cortices and provides for development of complex human intelligence.[32] New information enters the brain and connects with established neuron patterns by adapting them to new input.[33] Each time the new information is present, it will excite the reconstructed memory.

Environmental input is multisensory, and experience sensitizes receptivity.[34] A technique for recording magnetic brain fields, magnetoencephalography (MEG), was used to investigate how audio-tactile perception is integrated in brains of musicians and nonmusicians. The audio-tactile aspects of musical instrument playing was mimicked combining tones and tactile finger stimulation in a finger-to-tone relationship. Activation of brain networks responsible for audio-tactile integration demonstrated greater influence among those with musical training.[35] In a reading comprehension study using fMRI, researchers learned subjects' comprehension relied on the coactivation of the semantic control network (connected brain areas representing related meanings in text), and brain areas associated with mental model updating and integration.[36] Both results demonstrated the brain's associative functionality.

Classroom Applications

Four fourth-grade classes were studied as they learned about states of matter. In the experimental treatment classes, students solved problems, explained results, extended new knowledge to other situations, and expanded their understanding with concept maps and other devices. Traditional teaching methods such as reading texts, instructor lectures, and taking notes were used in control classes. In conclusion, posttest mind maps created by the experimental classes showed more associations between concepts and between words and concepts than in maps created by the control students.[37]

Movement and Dance Connections

Dancers learn movements by associating sequences into chunks and imprinting phrases into memory as grouped units. In Western dance traditions, movements are associated with counts, but dancers often remember movements by relating them to musical characteristics, spatial relationships, or their meaning.

Rehearsals help dancers associate actions with trigger stimuli. Professional musicians often have one rehearsal before a performance because they can read the music from associated notation. Dancers require a longer time to embody repertoire. Choreography is usually sculpted on dancers who perform it, so creating or reconstructing a work can take months. However, a lifetime later, upon hearing a familiar score, dancers often recall associated movements.

Associations define dance genres. These include sexual innuendos in jazz dance, images of weightlessness in ballet, Graham's archetypal symbolism, or hip hop's aggressive attitude. Spatial relationships trigger associated meanings—dancers positioned in a circle suggest community, while two lines of dancers coming toward one another impart confrontation, as in *West Side Story*'s fight scene.

Dance often uses associations based on gestures, shapes, designs, or relationships. As mentioned, Graham performed rocking motions in her role as a pioneer young bride/mother in *Appalachian Spring* to indicate holding and soothing a baby. Each member of an audience brings their own associations to a performance, leading to different interpretations for individual viewers. Strong associations wire movements firmly and produce clear memories of a dance.

Movement Explorations

Music Associations: (a) Listen to a piece of music you enjoy. (b) Create an image that springs to mind from the music. (c) Develop movement that relates to the image and music. (d) Practice the movement until it becomes automatic. (e) Wait a few days and play the music again. Does the movement come back into your body?

Spatial Associations: (a) Choose an area or space in your home, studio, or outside. (b) Create a simple movement phrase in that space. (c) Repeat the phrase several times until it is performed easily. (d) When you return to that space at a later time, does memory of the phrase return?

Symbolic Associations: (a) Think about associations connected with a stop sign and choose one. (b) Create a phrase of movement that communicates your association. (c) Practice the phrase. (d) Does the phrase come back to mind when you see a stop sign?

SIMILE, METAPHOR, AND ANALOGY

Simile and metaphor connect associations in language through comparisons that integrate similarities between separate entities. They add color to language and paint a robust mental picture. Analogy is another language-based association used to compare and describe similarities between the features or structure of different things. An example is comparing arteries to a stream's tributaries due to their structural similarity.

Brain Network

A quantitative meta-analysis was conducted of fMRI studies to determine the validity of an hypothesis that analogy and metaphor are linked to creativity and consistently activate the same brain systems. Two of the meta-analyses completely supported the hypothesis, but a third cluster of studies revealed analogy and metaphor share only some common neural systems.[38]

Classroom Applications

A seventh-grade Australian English teacher pushed his students to a deeper understanding of their texts. His first step was to have students connect text with visual images that served as a metaphor. Some of the metaphorical symbols selected by the students were literal, and others were more complex. Later, the students used a second text in which they connected random objects with parts of the text as interpretative images. The teacher believed the students learned to create rich metaphors by connecting text with features of objects.[39]

Movement and Dance Connections

Simile, metaphor, and analogy can be generalized under the term *metaphor*, as they all unify characteristics or functions of diverse entities. Metaphor enables audiences to derive meaning from abstract movement. It is a form of creative expression not limited to language but is centrally based.[40] Metaphor is internal to movement, and it gives rise to an association internal to the viewer.

Dance teachers use metaphor to improve instruction. Jumps can be heightened with a description of a bounding deer. Different body shapes, movements of various sizes, and energy qualities can be used as metaphors for teaching. For example, the dynamic in figure 8.1 is a metaphor for counterbalancing.

Metaphor is inherent in dance performance, even in abstract works. Spatial metaphors reflect life relationships: coming together suggests union, splitting apart infers rejection, and separateness reflects individuality. Metaphors become interdependent in developing meaning.

Figure 8.1 Counterbalancing. *Source*: Reprinted by permission, from S. C. Minton, 2007, *Choreography: A Basic Approach Using Improvisation*, 3rd ed. (Champaign, IL: Human Kinetics). Photographer Joe Clithero of B & J Creative Photography.

Movement Explorations

Simile: (a) Develop a written simile. (b) Create movement to express it. (c) Perform the movement simile for someone else. (d) Do they understand what your simile is?

Metaphor: (a) Create a movement phrase that communicates an idea or has meaning. (b) Develop a verbal metaphor for it. (c) Think about the metaphor and communicate it through movement. (d) Does dancing the metaphor change your original choreography?

Analogy: (a) Create a movement phrase with very fast actions that abruptly change directions and pathways. (b) Can you make verbal analogies to real-life situations based on the characteristics or functions of the movements?

TRANSFERENCE OF LEARNING

The cognitive process of transference, also called transformation, conveys knowledge from one discipline or medium to another form. There has been national dialogue about whether arts should be taught in schools for their ability to transfer knowledge into learning, or for intrinsic purposes of art for art's sake. Dance has incentive to enter education through the door of arts integration. Only 7.6 percent of public schools in America have dance programs, but integrated arts could provide an effective entrée.

Brain Network

Intelligence includes creating accurate connections between new information and existing neural patterns. With increased educational experiences, students are more likely to construct connections between new input and existing neural networks.[41] Experience and mental manipulation of input extend, change, and strengthen connections that contribute to intelligence.

When experiencing or learning something new, the brain searches for an existing neural network where new information will fit. If the fit is a good, positive transfer takes place. If new content is similar in some respects but not others, the fit is not complete, and transfer cannot occur.[42]

Classroom Applications

Transferring information to new contexts depends on a number of factors. A threshold of learning must be achieved in both mediums for transfer to occur. This means students must understand the connection between the mediums being used for transfer to take place. In the project *Science with Dance in Mind*, students learned about photosynthesis by physically dancing its qualities. Kinesthetically based experiences require understanding movement as well as learning a topic in a different discipline.[43] Positive transference depends on how similar the learning tasks are to one another.

Near transference is required as opposed to far transference. Learning in one discipline or medium must directly translate to learning in another. Thus, spatial designs in dance can resemble those in geometry and facilitate learning. The shapes of letters could be taught using body shapes. Far transference applies aspects of learning that are loosely related but do not directly apply to another discipline. Learning movement sequences does not directly teach reading or math.[44]

Transference has also been categorized as literal—a replication or imitation of some feature—or abstract—a general interpretation of a quality. Interpreting a drawing's jagged shape with sharp sounds or movements is an abstraction.[45] Realistic paintings are literal[46]; abstract art reveals properties or emotions and can be minimalist.[47]

Movement and Dance Connections

Dance transfers ideas into actions. Movement reveals the essence of knowledge, and students can actively and enjoyably transfer their understandings to memory through dance. Most topics can be danced. The Earth's atmosphere is understood by changing the level and quality of movement to represent the different atmospheric layers. The meaning of different words such as *accelerate* and *lethargic* can be transferred into experiential qualitative actions by dancing their meanings.[48]

A social studies project with fourth-through-eighth-graders in Washington, DC, public schools explored the cultures of a variety of countries. Initially, students brainstormed aspects comprising a culture, then checked off each one that affected how a society would dance, which inevitably included every item on the list. Small breakout groups created an imaginary society, defined the hallmarks of its culture, choreographed a dance its peoples would do, and performed it for classmates with no explanation. Classmates interpreted the values, beliefs, and customs of the culture. They learned to "read" movement and understand dance as an expression of values and beliefs in their curricular study of world countries.[49]

Mime or pantomime is a literal movement form dating back to the ancient Greeks. The Romans were so fond of pantomime it became popular entertainment throughout the empire.[50] Pantomime can be a literal expression used as transference of information, but is this dance? Classical ballet sometimes uses pantomime, but as an art form dance abstracts from realism and presents the essence of an idea. At moments, dance might incorporate literal gesture, pedestrian movement, or realistic body shapes, but this is within abstract structure and interpretation.

Movement Explorations

Literal Transference: (a) Think of an everyday activity such as putting on a pair of pants. (b) Pantomime the movements that you perform to accomplish the activity.

Abstract Transference: (a) Study a painting. (b) Choose one aspect in the painting and focus on its essence. (c) What basic feeling or feelings are communicated by its essence? (d) Transfer this essence to movement.

Transferring Content: (a) Draw numbers along a line to make a number line on the floor from one to twenty with the numbers one step apart. This can be done by placing numbers on a roll of paper towels. (b) As you go forward, the numbers get greater as you are adding. As you go backward, the numbers decrease and you are subtracting. (c) Practice adding and subtracting by counting different movements along the number line. (d) Apply the concepts of "more" in addition as numbers grow, and "less" in subtraction as numbers are reduced.

SYMBOL MAKING

Symbolism is the hallmark of the human mind, and is found in all cultures throughout history. It forms the basis of a collective communication and the backbone of language. A letter is a symbol representing a sound, and sequential sounds form words. The arts are symbolic media.

Brain Network

Alexander Luria wrote about a patient who had a bullet lodged in the left hemisphere at the juncture of the temporal, occipital, and parietal lobes. The temporal lobe processes sound and language, the occipital lobe visual images, the parietal lobe spatial relationships and integration of different sensory information, while the juncture between the areas produces sensory associations. The patient could not understand grammar dealing with relationships, comprehend whole words and sentences, or recall whole memories, since these cognitive functions draw relationships between symbols.[51]

Language and mathematics are composed of symbols. Brain images have shown an overlapping function of Wernike's and Broca's areas between the two disciplines. These areas are central to understanding and using symbols.[52]

The brain areas responsible for language and mathematics are also bilateral, meaning both hemispheres of the brain are active. The left side discerns details such as the exact meaning or sequence of symbols. The right side determines how to use both types of symbols in their appropriate context.[53]

Classroom Applications

Education could be viewed as a system that teaches skillful and informed use of symbols. Symbols are the backbone of communication. Each discipline has its own set of symbols that trigger memory cues and convey knowledge in succinct form.[54] Children learn at an early age that a plus sign represents addition, a minus sign means subtraction, and an equal sign symbolizes balance.

Symbols can persuade and transmit ideas, but their meaning must be matched to reality for effective communication. To accelerate learning, schools often teach symbols first without connecting them to realistic objects or scenes. Rather than accelerate learning, abstracting symbols from meaningful contexts causes confusion.[55]

Some symbols are archetypal and universal across cultures. Symbols of motherhood, sexuality, fertility, or death are understood in every society. Others are unique to a particular society, religion, or philosophy, such as the Christian cross or yin/yang symbol.

There are a myriad of symbols in daily life: traffic signs, street crossings, restrooms, no smoking, flags, political parties, even types of money. Foreigners learn to recognize cultural symbols far more quickly than learning the language.

The symbols of language or writing can affect brain use. Linear alphabets such as those in English and Russian Cyrillic flow linearly from left to right, while Arabic or Hebrew progress from right to left. Asian languages use picture characters instead of letters and promote holistic thinking rather than linear progression.

Cognitively, symbol making is a high-level thought process, requiring a broad understanding usually learned from parents or cultural indoctrination. Cognitive scientists differentiate between a symbol and a sign. Signs have a specific and literal meaning, hence are concrete images or metaphors. A stop sign requires a specific action. It is believed the ability to abstract develops in adolescence. A car, which in childhood is connected to its function, becomes a symbol of independence and status.

Movement and Dance Connections

Symbols are more global than images or metaphors. Symbolic movements unify "collective" significance in relation to human nature. The movements of a distressed or angry infant are jerky and aggressive with the hands usually curled into fists. It follows that a shaking fist is understood as a universal symbol of anger or aggression, while a smile or fluid wave is welcoming or friendly. For infants thirteen-to-sixteen months, when a word is spoken in contradiction to a gesture the word will be ignored.[56] "Such speech-gesture conflicts can negatively affect motor performance."[57]

Dance is symbolic movement. Through the millennia, rituals were abstracted into purposeful symbols. The Lion Dance of the African Masai involves high jumps, in which their height demonstrates power, just as the high leaps of a ballet dancer portray strength and agility. In ritual dance and jazz, sensual hip undulations symbolize female fertility, while aggressive pelvic thrusts indicate the sexual act. Dance props, costumes, and scenery are also rife with symbolic images.

Dance is a universal symbolic language that extends beyond the boundaries of verbal language. The values and beliefs of a culture are revealed through the qualities and characteristics of its movement forms. Perhaps, through dance, people of different cultures can learn to understand one another, "wire together" on a deeper human level, and overcome some of the separations that cause conflicts.

Movement Explorations

Movement Symbols: (a) Think of as many symbolic movements or gestures as you can. (b) Perform them.

Language Symbols: (a) Think of a word that communicates a particular meaning. (b) Analyze your feelings in relation to the meaning of the word. (c) Create a movement based on this meaning.

Domain and Discipline Symbols: (a) Find a symbol that is associated with a particular domain or discipline. (b) Research the meaning of this symbol. (c) Create a movement sequence that communicates the meaning of the symbol you selected.

Cultural Symbols: (a) Choose a symbol that is associated with a specific group or culture or one that is universal across cultures. (b) Research the meaning of the symbol you selected. (c) Create a sequence of movements that communicate this meaning.

NOTES

1. Eric Jensen, *Arts with the Brain in Mind* (Alexandria, VA: Association for Supervision and Curriculum Development, 2001).

2. John Dewey, *Schools of Tomorrow* (New York, E. T. Dutton, 1915); John Dewey, *Democracy and Education* (New York: Macmillan, 1916).

3. "The Quotations Page" accessed June 10, 2015, http://www.quotationspage.com/quotes/Confucious.

4. John Dewey, *Experience and Education* (New York, Macmillan, 1938).

5. Ron Ritchhart, Mark Church, and Karin Morrison, *Making Thinking Visible: How to Promote Engagement, Understanding, and Independence for All Learners* (San Francisco: Jossey-Bass, 2011).

6. Linda Darling-Hammond, *The Right to Learn: A Blueprint for Creating Schools That Work* (San Francisco: Jossey-Bass, 1997).

7. David Perkins, "From Idea to Action," in *The Project Zero Classroom: Views on Understanding*, eds. Lois Hettland and Shirley Veenema (Cambridge, 1999), 17–25.

8. National Research Council, *How People Learn: Brain, Mind, Experience, and School* (Washington, DC: National Academy Press, 2000).

9. Rima Faber, "Final Report for Science with Dance in Mind" (report, Baltimore County Public Schools, Baltimore, 2013).

10. Alison M. Rhodes, "Dance in Interdisciplinary Teaching and Learning," *Journal of Dance Education* 6, no. 2 (2006): 48–56.

11. Carl Flink and David J. Odde, "Science + Dance = Bodystorming," *Trends in Cell Biology* 22, no. 12 (2012): 613–16.

12. "University Studies," accessed March 5, 2015, http://www.jmu.edu/universitystudies/danceofAS.html.

13. Glenna Batson with Margaret Wilson, *Body and Mind in Motion: Dance and Neuroscience in Conversation* (Bristol, UK/Chicago: intellect, 2014).

14. Kurt W. Fischer and Katie Heikkinen, "The Future of Educational Neuroscience," in *Mind, Brain, & Education: Neuroscience Implications for the Classroom*, ed. David A. Sousa (Bloomington, IN: Solution Tree Press, 2010), 249–69.

15. John J. Ratey with Eric Hageman, *Spark: The Revolutionary New Science of Exercise and the Brain* (New York: Little Brown, 2008).

16. Beatriz Calvo-Merino, "Neural Mechanisms for Seeing Dance," in *The Neurocognition of Dance: Mind, Movement and Motor Skills*, eds. Bettina Blasing, Martin Puttke, and Thomas Schack (New York: Psychology Press, 2012), 153–76.

17. Fischer and Heikkinen, "The Future of Educational Neuroscience"; H. Lynn Erickson, *Concept-Based Curriculum Instruction for the Thinking Classroom* (Thousand Oaks, CA: Corwin Press, 2007).

18. Robert and Michèle Root-Bernstein, *Sparks of Genius: The 13 Thinking Tools of the World's Most Creative People* (Boston: Houghton Mifflin, 1999).

19. Gillian Judson and Kieran Egan, "Engaging Students' Imaginations in Second Language Learning," *Studies in Second Language Learning and Teaching* 3, no. 3 (2013): 343–56.

20. Dale Rose, Susan McMahon, and Michaela Parks, "Basic Reading through Dance Program," *Evaluation Review* 27, no. 1 (2003): 101–25.

21. Steven White, "Factors That Contribute to Learning Differences among African American and Caucasian Students," 1992 ERIC ED 374177; Clara C. Park, "Learning Style Preferences of Asian Americans (Chinese, Filipino, Korean and Vietnamese Students in Secondary School," *Equity and Excellence in Education* 30, no. 2 (1997): 6–77; Park, "Learning Style Preferences of Korean-Mexican Armenian-American and Anglo Students in Secondary School," *National Association of Secondary School Principals Bulletin* 81, no. 585 (1997): 103–11.

22. Betsy Cooper, "Embodied Writing: A Tool for Teaching and Learning in Dance," *Journal of Dance Education* 11, no. 2 (2011): 53–59.

23. Elliot W. Eisner, *The Arts and the Creation of Mind* (New Haven, CT: Yale University Press, 2002).

24. Judy Willis, "The Current Impact of Neuroscience on Teaching and Learning," in *Mind, Brain, & Education: Neuroscience Implication for the Classroom*, ed. David A. Sousa (Bloomington, IN: Solution Tree Press, 2010), 45–66.

25. James E. Zull, *From Brain to Mind: Using Neuroscience to Guide Change in Education* (Sterling, VA: Stylus Publishing, 2011).

26. Zull, *From Brain to Mind*.
27. V. S. Ramachandran, *The Tell-Tale Brain: A Neuroscientist's Quest for What Makes Us Human* (New York: W. W. Norton, 2011).
28. Zull, *From Brain to Mind*.
29. Jacqueline McGinty, Jean Radin, and Karen Kaminski, "Brain-Friendly Teaching Supports Learning Transfer," *New Directions for Adult and Continuing Education* 137 (Spring 2013): 49–59.
30. Oliver Sacks, *Musicophilia* (New York: Random House, 2007).
31. Steven Brown and Lawrence M. Parsons, "The Neuroscience of Dance," *Scientific American* 299 (2008): 78–83.
32. Zull, *From Brain to Mind*, 86.
33. Eric Kandel, *In Search of Memory: The Emergence of a New Science of Mind* (New York: W. W. Norton, 2006).
34. Anja Kuchenbuch, Evangelos Paraskevopoulos, Sibylle C. Herholz, and Christo Pantev, "Audio-Tactile Integration and the Influence of Musical Training," *PLoS One* 9, no. 1 (2014): 1–15.
35. Kuchenbuch, Paraskevopoulos, Herholz, and Pantev, "Audio-Tactile Integration and the Influence of Musical Training."
36. Katherine Swett, Amanda C. Miller, Scott Burns, Fumiko Hoeft, Nicole Davis, Stephen A. Petrill, and Laurie E. Cutting, "Comprehending Expository Texts: The Dynamic Neurobiological Correlates of Building a Coherent Text Representation," *Frontiers in Human Neuroscience* 7 (2013): 1–14.
37. Ibrahim Bilgin, Humeyra Coskun, and Idris Aktas, "The Effect of the 5E Learning Cycle on Mental Ability of Elementary Students," *Journal of Baltic Science Education* 12, no. 5 (2013): 592–607.
38. Oshin Vartanian, "Dissociable Neural Systems for Analogy and Metaphor: Implications for the Neuroscience of Creativity," *British Journal of Psychology* 103 (2012): 302–16.
39. Ritchhart, Church, and Morrison, *Making Thinking Visible*.
40. Jay Seitz, "The Development of Metaphoric Understanding: Implications for a Theory of Creativity," *Creativity Research Journal* 10, no. 4 (1997): 347–53.
41. Willis, *Mind, Brain, & Education*.
42. Patricia Wolfe, *Brain Matters: Translating Research into Classroom Practice* (Alexandria, VA: Association for Supervision and Curriculum Development, 2001).
43. Faber, "Final Report for Science with Dance in Mind."
44. Richard J. Deasy, ed. *Critical Links: Learning in the Arts and Student Academic and Social Development* (Washington, DC: Arts Education Partnership, 2002); James E. Catterall, "The Arts and the Transfer of Learning," in *Critical Links*.
45. Ramachandran, *The Tell-Tale Brain*.
46. Eisner, *The Arts and the Creation of Mind*.
47. Root-Bernsteins, *Sparks of Genius*.
48. Minton, *Using Movement to Teach Academics: The Mind and Body As One Entity* (Lanham, MD: Rowman & Littlefield Education, 2008).
49. Rima Faber, Projects in District of Columbia Public Schools, 1980–1997.
50. Gayle Kassing, *History of Dance: An Interactive Arts Approach* (Champaign, IL: Human Kinetics, 2007).

51. Norman Doidge, *The Brain That Changes Itself: Stories of Personal Triumph from the Frontiers of Brain Science* (New York: Penguin, 2007).

52. Zull, *From Brain to Mind*.

53. Zull, *From Brain to Mind*.

54. Zull, *From Brain to Mind*; Mihaly Csikszentmihalyi, *Creativity: Flow and the Psychology of Discovery and Invention* (New York: Harper Perennial, 1996).

55. Zull, *From Brain to Mind*.

56. Jay Seitz, "The Body Basis of Thought," *New Ideas in Psychology* 18 (2000): 23–40.

57. Seitz, "The Body Basis of Thought," 31.

Chapter 9

Problem Solving

Since the era of ancient Greece, when Socrates refined a pedagogical methodology based on posing questions, students have been solving academic problems. Howard Gardner defines intelligence as an ability to solve practical problems; a realistic intelligence that enables humans to adapt. To solve problems requires logic, flexibility, and imagination.[1] It fulfills intent, furthers knowledge, or inspires creative thinking to achieve goals. The brain is a natural problem solver that seeks novel methods.

CREATIVE PROBLEM SOLVING

Problem solving is a form of creative active learning. Students discover meanings with a sense of achievement and satisfaction. In this discussion, creative thinking is not differentiated from creative problem solving since it is the basis of problem solving. The problem is a stimulant for creativity.

Brain Network

Solving problems requires activity in both cerebral hemispheres, but the path to solutions lies through the corpus callosum. The most significant creative activities in human culture were made possible through collaboration of the left and right brain hemispheres.[2]

Thinking creatively engages divergent thinking, which summons a variety of cognitive functions activating diverse brain regions. Neuroscientists agree divergent thinking involves coactivation and communication among brain regions normally not connected during other types of thought. It combines a high level of knowledge in specialized areas, mediated by frontal lobe processing during the use of working memory and sustained attention.[3] When a problem is successfully solved, the brain produces dopamine as a reward, which encourages further work.[4]

Using EEGs, a study found more active brain areas during creative cognition than in conventional thinking. Other studies matched scores on the Torrance Test of Creative Thinking (TTCT) with measurements of cerebral blood flow and found subjects with high scores on the TTCT, in comparison to students with average scores, had significantly greater activity in brain structures responsible for cognition, emotion, working memory, and response novelty. An unexpected result found patients with frontal lobe damage experienced a lack of inhibition with bursts of activity across brain regions.[5]

A study on transfer of learning hypothesized that performing arts students would perform differently on the Uses of Object Task than nonperforming arts students, and would show different brain activation patterns. In this task, a measure of divergent thinking, subjects are given an object like a brick and asked to think of multiple uses. The fMRI results showed the performing arts students displayed activation patterns in the left inferior gyrus and left superior frontal gyrus, and the nonperforming arts students did not. The brain areas activated in the performing arts students are involved in language processing.[6]

Imaginative imagery leads the mind in new directions, hence it promotes creative problem solving. Images are manipulated through the frontal cortex and working memory but must be congruent to the problem to be helpful.

The neurotransmitter acetylcholine stimulates attention, learning, and enhances memory. Research showed rats trained on difficult spatial problems have a higher percentage of acetylcholine in their brains than rats solving simple spatial issues. Other animal experiments demonstrated that a challenging or enriched environment increases brain weight in the cerebral cortex because neurons grow, develop branches, and make more connections with other neurons. These can be lifelong changes.[7]

Classroom Applications

A problem must be defined, information gathered, and possibilities explored before it can be solved. This process teaches students to read pertinent subject matter, do research, interview knowledgeable individuals, and become immersed in content. Sir James Lighthill exemplified a creative approach when he researched the aerodynamic conditions for flight by swimming in a variety of conditions. He learned the behavior of waves, tides, and currents of water that he applied to the necessary conditions for flight.[8]

In creative problem solving, the criteria are spelled out, but the actual form of the solution is left open.[9] The arts offer open-ended solutions versus single correct answers. An artist begins with an inspiration or intent but may have no distinct idea of how the artistic product will take shape. Immersion in information surrounding a problem is an integrative process that can give rise to sudden insight—a moment known as the "aha" experience.[10]

Group collaborations are pedagogically effective. Collective discussions with social interactions contribute to heightened engagement among students and a wider range of possible solutions. Students deepen understanding of concepts, gain skills from others, and experience excitement from achieving a solution.[11]

Movement and Dance Connections

Dance instructors explore diverse methods to interest and effectively teach all students in a class. Students face difficult physical and personal issues adapting their bodies to dance technique, but they can achieve technical skill through creative problem solving by applying multiple methods to execute, synthesize, assess, and select the best solutions, and practice to achieve excellence.

Choreographers manipulate, weave, vary, and ultimately select movements that communicate an artistic intent and shape a unified dance. Performers internalize and fulfill the choreographer's artistic intent. To bring a performance to life, they recreate a work to make it their own. Otherwise the performance is imitative and lacks meaning. In order to produce their work, choreographers solve problems concerning lighting, costuming, music, and programing, while keeping the intent of the dances and the audience perspective in mind.

Movement Explorations

Define Problem (Intent): (a) Imagine that your intent is to travel through a space, first in a pleasant environment (beautiful field of flowers); then an unpleasant environment (lightning storm); and finishing back in your pleasant environment. (b) How would you communicate each of these environments? What qualities of movement would be used to respond to each of these environments? (c) What actions would happen in each of these environments? (d) How would you make these actions abstract instead of pantomime?

Explore (Improvise): Improvise freely to find movements that respond to these environments.

Synthesis: Plan a method of organizing the movements you created into a short dance with a beginning, middle, and conclusion.

LOGICAL PROBLEM SOLVING

Logic is commonly delineated as deductive reasoning, inductive reasoning, and inference.[12] Deductive reasoning is a process of deconstruction in which a general concept is related to specific instances; a process used in basic mathematics. Inductive reasoning is constructive thinking, whereby specific instances are related to determine a single concept or phenomena—a process applicable to scientific hypotheses. Inference takes place when one concept or phenomena suggests another. Logic in the arts is inference.

Brain Network

Executive functions enable humans to logically reason and solve problems.[13] They are the brain's cognitive controls and attentional system that directs working memory, reasoning, task flexibility, and problem solving. Executive functions are regulated by the prefrontal region of the frontal lobes in coordination with other brain areas depending on the task. Executive skills are the product of executive functions: attending, inhibiting, modulating, planning, organizing, and associating.[14]

Classroom Applications

The steps in integrative, logical problem solving are: (1) make a plan to solve the problem. The better the plan, the more likely a solution will be found; (2) choose memories and facts essential to the plan; (3) manipulate the necessary sequence of actions; and (4) test the solution in practice. Logical problem solving is an active, constructive process in which the purpose can change, facts can be discarded, thinking can be altered, and elements can be rearranged or added in order to fit the goal of the problem.[15]

The knowledge necessary for solving a problem must be recalled from short- and long-term memory to be useful in finding a solution. Bits of knowledge are manipulated in short-term memory to provide possibilities of organizing relevant content. Each method is tested conceptually, physically, or on a practical level to determine which provides the most viable solution.

In traditional pedagogy for a subject such as math, students use a specific formula demonstrated by the teacher to find an answer. In contrast, when allowed to figure out how to solve a problem themselves, students are more interested, stimulated, and motivated.[16]

Movement and Dance Connections

The dancing body is an instrument for artistic movement. It is dependent on anatomy, adheres to the laws of physics, and thus functions logically.

To learn technique, a dancer must understand kinesiology in relation to gravity. It requires analysis and logical thought. Admittedly, many dance teachers and dancers do not receive training in anatomy or kinesiology and therefore misuse and often abuse bodies. Consequently, gifted dancers are injured and forced to end careers early.

Choreography embodies logical problem solving as well as creativity. Liz Lerman adheres to a Logic Model for problem solving. She chooses a topic for her work, usually involving an issue of concern. Before she rehearses with dancers, she researches and gathers information about the topic, plans sections of a dance, often prepares a spoken script, and develops improvisational structures for her cast from which to choose movements.

Rudolf Laban, choreographer and renowned movement theorist, developed logical systems of movement analysis for performing, observing, and describing movement. He created movement choruses and choreographic structures shaped by improvisational themes. Laban's dancers had to be skilled collaborators, sensitive to each other, and trained in choreutics (the study of spatial forms of movement) and eukinetics (the study of rhythm and dynamics of movement).[17]

A notation system in the arts combines linear cognition with holistic brain use. For almost two thousand years, music has had a system of notation that communicates repertory. Notation was spottily attempted in dance, but none took hold. Laban solved the issue by inventing Labanotation, a notation system using a graphic set of symbols intuitively shaped or shaded to indicate directions, levels, positions of body parts, tempo, and movement dynamics. It was later perfected by Ann Hutchinson Guest and is currently the most widely accepted system.[18]

Movement Explorations

Define the Problem: Decide on a topic or movement issue to be explored as a problem.

Plan: (a) What will you do to find movements? Will you develop the movements for each part and then go on to the next, or will you follow your inspiration? (b) How will you order your movements?

Explore: Complete your plan by exploring and experimenting to develop movement possibilities.

Select: (a) Select the movements from your exploration that fulfill your plan. (b) Arrange the movements in a sequence.

Solution Tested in Practice: (a) Perform the sequence or arrangement you created without stopping between each separate action. (b) Make necessary adjustments for smooth transitions between movements.

THE SOLUTION

Creative problem solving pursued through exploration often leads to several alternative solutions.[19] It rarely proceeds forward in a linear progression because the final stage of evaluation and elaboration often loops back to the beginning in a revision process.[20] Scientific methodology defines a problem by testing an hypothesis, examining evidence, looking for patterns, raising questions, and challenging findings. Creative transformational thinking is a synthesis of understanding. Problem solving requires perceiving parts in relation to the whole.[21]

Brain Network

Neural impulses move in all directions: up and down a vertical axis between the brain stem and cortex, from left to right, and from back to front. This enables the brain to synthesize information and assemble it into global relationships. All of these processes along with the circulation of blood and neurochemicals take place simultaneously in multiple regions of the brain.[22] The release of the neurotransmitter dopamine upon finding a successful solution produces pleasure.

Classroom Applications

Teachers often expect students to arrive at one correct solution rather than seek a variety of possibilities. Fifty subjects watched thirty-four videos of magic tricks and were instructed to figure out how the tricks were done. The participants also had to indicate if they used insight or some other method to understand a trick. After fourteen days, participants recalled how the tricks were accomplished. It was discovered that 64.4 percent of intuitive solutions remembered correctly in comparison to only a 52.4 percent of consciously processed solutions.[23] Perhaps an intuitive "aha" insight is the more effective method of solving problems and learning.

Movement and Dance Connections

As mentioned, problem solving does not always proceed forward in a linear fashion. When discussing their dance-making process, one member of the group Pilobolus said, "Then there might also be phrases that don't seem to have any home. The dance might go fine A to F, then suddenly you've got M-N-O, and you have to then think of an idea that would make a legitimate connection theatrically."[24]

A qualitative research approach was used to examine the cognitive process of fifth-graders during dance making. The children's perceptions of their process was categorized by their meanings. It was found that generative processes are complex combinations that form a mental synthesis of new structures. Attention to structure was seen at almost every stage of the children's dance-making process.²⁵

Movement Explorations

Creative Solutions: (a) Look at the dancers in figure 9.1. They are poised to continue moving. (b) What is your hypothesis concerning the next series of

Figure 9.1 The dancers are poised to continue moving. *Source*: Reprinted by permission, from S.C. Minton, 2007, *Choreography: A Basic Approach Using Improvisation*, 3rd ed. (Champaign, IL: Human Kinetics). Photographer Joe Clithero of B & J Creative Photography.

movements to be performed by the dancers? (c) How did you deduce your response? (d) What movements do you think could follow from the image shown?

Scientific Solutions: (a) Working with a partner, find as many possibilities to balance together as you can. (b) What made the balances work?

Insight: (a) Analyze the thinking process you used in each of the previous explorations. (b) Did you experience a sudden insight in solving either of these two explorations?

REFLECTION

During reflection, students evaluate their solution to a problem and, perhaps, extend new knowledge to a broader spectrum of content and experiences. Many teachers use a self-reflective, metacognitive strategy in which they encourage students to think about prior knowledge, what they learned, what they want to learn, and why they think so. Thinking maps and well-constructed questions also encourage reflective thought.[26] These devices encourage students to understand what they do and why.

Brain Network

Reflection involves both subconscious and conscious brain processes. Even when the brain is not being stimulated by outside sensory information, its neurons are active. During subconscious neuronal activity, pathways between neurons resembled those found during actual virtual experiences. In the absence of new experiences, the brain replays experiences from the past.[27]

Conscious reflection revisits previous neural connections in the cortex and wanders among them. Mental explorations connect information learned previously with meanings. Reflection involves many of the neural processes discussed previously—memory formation, recall, chunking, consolidation, and reconsolidation of memories.

Classroom Applications

The reflective teacher or student uses ongoing metacognitive activity.[28] Successful problem solving is enhanced by metacognitive awareness of one's learning or teaching and to expand processes and preferences in service of the problem to be solved.[29]

In a college education course, a first assignment asked prospective teachers to reflect on and describe their own learning experiences—one that was successful and another that was not. The students analyzed key characteristics of a successful experience. They used metacognitive processes to reflect on and analyze their own learning in order to understand what constitutes a good educational situation.[30]

Reflection keeps the critical process on track by determining the clarity, accuracy, consistency, relevance, depth, and breadth of creative work as it relates to preset goals.[31] Stepping back in thought clarifies what is known.[32] Reflection has been used in reading classes to help students determine if they met the purpose of an assignment and were able to integrate new learning with prior knowledge. Teaching tools such as time lines, flow charts, graphs, tables, mind maps, Venn diagrams, and outlines aid reflection.[33]

Teachers encourage reflective thinking by asking questions that stimulate focus. Conceptual questions are particularly effective to identify patterns and transfer understandings.[34] Examples of factual and conceptual questions can be found in chapter 3, page 42.

Movement and Dance Connections

Reflection is ongoing for learning dance technique and dance making. Dancers use reflection to stimulate body awareness and correct technical problems. Understanding of movement is fundamental to developing dance skills. Feedback, assessment, and technology in the form of video recordings are tools used to enhance students' reflective practice. The reflective dance teacher analyzes the effectiveness of instruction to evaluate and plan necessary changes.[35]

Choreographers use reflection to assess and revise their work. Following reflection, the solution might be to: (1) create new movements; (2) change existing movements; (3) provide a better transition between parts of a dance; or (4) change the structure of the dance. The possibilities are infinite.

Reflection provides analysis of dance performances whether feedback is provided verbally or written as a review or critique. "During reflection, students have an opportunity to clarify and record their impressions of the structures and qualities they perceived in a work before giving the choreographer any verbal response . . . or formulating interpretations and judgments."[36] It is helpful for students to journal as reflective assessment practice.

Movement Explorations

Reflective Technique: (a) Perform a jump or a turn. (b) Reflect on your performance. (c) What would improve your performance? If it was performed skillfully, what made it so?

Reflective Response: (a) Reexamine some explorations performed toward the beginning of the book. (b) Reflect on information learned since performing this early exploration. (c) How would you apply your current level of understanding to your solution for earlier exploration?

Factual Thinking: The rainforest has thick, dense foliage. Create movement you might use to push through the dense foliage.

NOTES

1. Howard Gardner, *Frames of Mind: The Theory of Multiple Intelligences* (New York: Basic Books, 1983).
2. Carl Sagan, *The Dragons of Eden* (New York: Random House, 1977), 191–95.
3. Mariale M. Hardiman, "The Creative-Artistic Brain," in *Mind, Brain, & Education: Neuroscience Implications for the Classroom*, ed. David A. Sousa (Bloomington, IN: Solution Tree Press, 2010): 227–46.
4. Daniel T. Willingham, *Why Don't Students Like School? A Cognitive Scientist Answers Questions about How the Mind Works and What It Means for the Classroom* (San Francisco: Jossey-Bass, 2009).
5. Hardiman, "The Creative-Artistic Brain."
6. Kevin Niall Dunbar, "Arts Education, the Brain and Language," in *Learning, Arts, and the Brain: The Dana Consortium Report on Arts and Cognition*, eds. Carolyn Asbury and Barbara Rich (New York/Washington, DC: Dana Press, 2008), 81–91.
7. Norman Doidge, *The Brain That Changes Itself: Stories of Personal Triumph from the Frontiers of Brain Science* (New York: Penguin Books, 2007).
8. Robert and Michèle Root-Bernstein, *Sparks of Genius: The 13 Thinking Tools of the World's Most Creative People* (New York: Houghton Mifflin, 1999).
9. Elliot W. Eisner, *The Arts and the Creation of Mind* (New Haven, CT: Yale University Press, 2002).
10. Mihaly Csikszentmihalyi, *Creativity: Flow and the Psychology of Discovery and Invention* (New York: Harper Perennial, 1996).
11. Jennifer Holm, "Perceptions of Problem Solving in Elementary Curriculum," *delta-K* 50, no. 2 (2013): 36–42.
12. "Logical Reasoning," *Wikipedia*, accessed December 18, 2014, http://en.wikipedia.org/wiki/logical_reasoning.
13. "Cognitive Functions," *Happy Neuron*, accessed October 23, 2014, http://www.happy-neuron.com/brain-and-training/executive-functions.
14. George McCloskey, "Executive Functions in Classrooms" (Learning and the Brain lecture, professional development seminar, University of Maryland, College Park, MD, November 17, 2014).
15. James E. Zull, *From Brain to Mind: Using Neuroscience to Guide Change in Education* (Sterling, VA: Stylus Publishing, 2011).
16. Daniel Pink, *Drive* (New York: Penguin Books, 2009).
17. Valerie Preston-Dunlap, *Rudolf Laban: An Extraordinary Life* (London: Dance Books, Ltd., 1998).
18. Ann Hutchinson Guest, *Labanotation: The System of Analyzing and Recording Movement* (New York: Theater Arts Books, 1954); Ann Hutchinson Guest, *Your*

Move: A New Approach to the Study of Movement and Dance (Luxembourg: Gordon and Breach Publishers, 1995).

19. Michael Michalko, *Cracking Creativity: The Secrets of Creative Genius* (Berkeley, CA: Ten Speed Press, 2001); H. Lynn Erickson, *Concept-Based Curriculum Instruction for the Thinking Classroom* (Thousand Oaks, CA: Corwin Press, 2007).

20. Csikszentmihalyi, *Creativity*.

21. Mark Reardon and Seth Derner, *Strategies for Great Teaching: Maximize Learning Moments* (Chicago: Zephyr Press, 2004).

22. Reardon and Derner, *Strategies for Great Teaching*.

23. Amory H. Danek, Thomas Fraps, Albrecht von Muller, Benedikt Grothe, and Michael Ollinger, "Aha! Experiences Leave a Mark: Recall of Insight Solutions," *Psychological Research* 77 (2013): 659–69.

24. Jean M. Brown, Naomi Mindlin, and Charles H. Woodford, eds., *The Vision of Modern Dance: In the Words of Its Creators*, 2nd ed. (Hightstown, NJ: Princeton Book Company, 1998), 172.

25. Miriam Giguere, "Dancing Thoughts: An Examination of Children's Cognition and Creative Process in Dance," *Research in Dance Education* 12, no. 1 (2011): 5–28.

26. Patricia Liotta Kolencik and Shelia A. Hillwig, *Encouraging Metacognition: Supporting Learners through Metacognitive Teaching Strategies* (New York: Peter Lang Publishing, 2011).

27. Zull, *From Brain to Mind*.

28. Erickson, *Concept-Based Curriculum Instruction for the Thinking Classroom*.

29. Kolencik and Hellwig, *Encouraging Metacognition*.

30. John Bransford, Sharon Derry, and David Berliner, "Theories of Learning and Their Roles in Teaching," in *Preparing Teachers for a Changing World: What Teachers Should Learn and Be Able to Do*, eds. Linda Darling-Hammond and John Bransford (San Francisco: Jossey-Bass, 2005), 40–87.

31. Erickson, *Concept-Based Curriculum Instruction for the Thinking Classroom*.

32. Ron Ritchhart, Mark Church, and Karin Morrison, *Making Thinking Visible: How to Promote Engagement, Understanding, and Independence for All Learners* (San Francisco: Jossey-Bass, 2011).

33. Kolencik and Hellwig, *Encouraging Metacognition*.

34. Erickson, *Concept-Based Curriculum Instruction for the Thinking Classroom*.

35. Lara Tembrioti and Niki Tsangaridou, "Reflective Practice in Dance: A Review of the Literature," *Research in Dance Education* 15, no. 1 (2014): 4–22.

36. Larry Lavender, *Dancers Talking Dance: Critical Evaluation in the Choreography Class* (Champaign, IL: Human Kinetics, 1996), 68.

Chapter 10

Twenty-First-Century Skills

In 1990 the first President Bush developed the Commission on Skills of the American Workforce, a committee comprised of leading educators, prominent business and government leaders, and scholars to study American education in relation to successful employment.[1] The Commission produced a report that predicted routine factory work would be outsourced to developing nations, and jobs in the United States would require education in technology. Focus on high-value products and services were recommended, along with education needed to produce workers who could compete globally.[2]

Sixteen years later, George W. Bush released a report of the New Commission on Skills of the American Workforce that alarmed the nation with its finding that American education ranked twenty-seventh out of fifty leading nations, far below expectations and behind some developing countries.[3] The report revealed that while the cost of education spiraled upward, employment of high school graduates had not improved since the 1970s. The jobs surviving in America were direct service or executive positions requiring creativity, problem solving, collaboration, and leadership skills.

Skills needed for successful employment were being ignored in an educational system built on testing. Factory jobs were outsourced to developing nations with cheap labor, and middle management was moved overseas to avoid high salaries and taxes. Only high-echelon positions and direct service jobs were available in American industry—jobs requiring innovation, problem-solving, decision-making, and management skills. The educational focus on test taking did not foster critical thinking, communication, collaborative, or creative skills that built America.

The Partnership for 21st Century Skills was born and defined thirteen skills in areas of: (1) Life and Career Skills, (2) Learning and Innovation Skills, (3) Information Media and Technology Skills, and (4) Core Subjects and

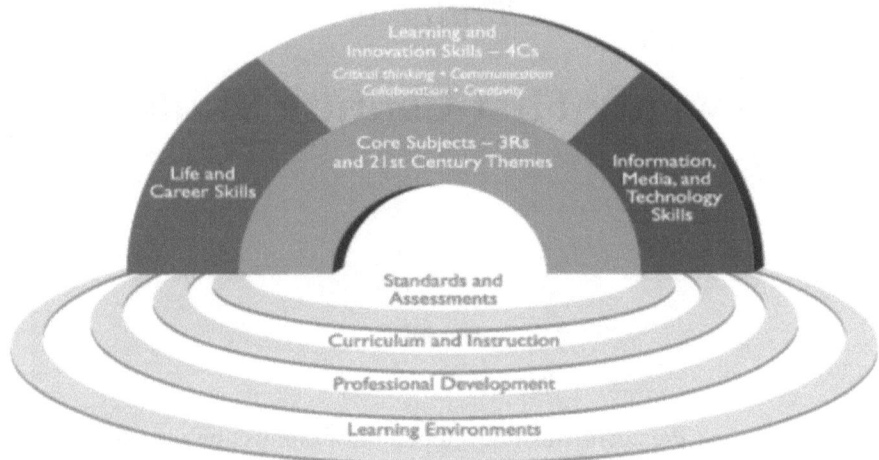

Figure 10.1   21st Century Student Outcomes and Support System. *Source*: The Partnership for 21st Century Learning (P21), www.P21.org.

21st Century Themes.[4] Innovation focuses on the four Cs: critical thinking, communication, collaboration, and creativity. The arts develop learning associated tasks, stimulate mental imagery, require effort over time, and provide cooperative opportunities—all skills necessary for success[5] (see figure 10.1).

CRITICAL-THINKING SKILLS

The field of neuro-education focuses on how children learn in relation to brain function. Studies show the neural networks engaged in creative thinking and problem solving are complex and that the arts can enhance cognition and learning.[6]

Brain Network

Critical-thinking skills involve executive functions, especially when task performance is expected. The frontal lobes direct other areas of the brain depending on how the task is to be accomplished. Working memory supplies awareness of information, the amygdala and limbic system contribute emotional content, the parietal area provides sensory input and activates motion, and the frontal lobe conducts orchestration and integration, producing successful interaction.[7]

Verbal communication involves left temporal lobe sites; schemata rely heavily on the right parietal cortex; and procedural knowledge is based in a number of brain structures—the premotor cortex, cerebellum, and the basal ganglia.[8] Research demonstrates the complexity of neural networks engaged in artistic creative thinking and problem solving, and that the arts promote cognition and learning.[9]

Classroom Applications

Ten low-income elementary schoolchildren were studied in an after-school arts program. The students painted a box on the inside to represent how they saw themselves, and on the outside to represent how others saw them. In another project, the students embellished juice containers to represent their ideal school. The children were encouraged to talk about the meaning of their artwork and view the work of others. A critical-thinking test given pre and post showed a significant increase in students' scores, a result that mirrored their improved ability to communicate with words and images.[10]

This demonstrates that inquiry-based arts classes effectively teach the twenty-first-century skills of creative innovation, problem solving, self-direction, initiative, productivity, responsibility, communication skills, and experienced flexibility, because no one project was exactly like those created by others.

Movement and Dance Connections

The deepest critical thinking occurs when interpreting, analyzing, critiquing, and relating dance to other experiences and learning.[11] These thought processes are highlighted most in creating dance, reflective performing, responding to dance, and connecting dance with personal experiences, contextual meaning, cultural significance, historical events, or other life phenomena. Imitating or repeating dance movements without reflection does not promote critical thinking.

The National Core Arts Standards in Dance were therefore categorized in the art-making processes of Creating, Performing, Responding, and Connecting. These processes occur simultaneously in practice, unlike the linear nature of language and logic. "The dancer imagines and improvises movements (creating), executes the movements (performing), reflects on them (responding), and connects the experience to other contexts of meaning or knowledge (connecting)."[12] One lesson can address all these processes by providing activities in which students create movement, think critically, communicate ideas, and relate movement to meaning.

The *National Core Arts Standards in Dance* incorporate twenty-first-century skills. Each standard begins with an age-appropriate cognitive verb that links critical thinking and artistic actions. The Pre-K-to-second-grade standards center on concrete, physical dance experiences. From third to fifth grade, focus is on how the body functions, movement skills, and the role of the elements of dance making. Middle school students are involved with relationships, and reflect on dance relationships and contrasts in dance making. High school students are wired to understand abstract concepts, and can address aesthetics, choreographic structures, and critical criteria.

Movement Explorations

Creating: Create your own movement sequence that includes at least five or six different actions.
Performing: Try performing these actions in several different orders.
Responding: Decide which order is the most fluid or organic.
Connecting: (a) Reflect on the movements you have chosen. (b) Do they have any connotations, symbolism, or connections to other meanings or experiences?

COMMUNICATION

Communicating well requires skill relating a person's reality to another person's experience. It transfers information through symbolic systems of words, gestures, images, or signs that have communal meaning. Communication is not a simple process since each individual has a set of life experiences that change their brain and reshape meanings. Symbols require cultural agreement. When driving a car, a red light means "stop" and a green light means "go." The word *mother* means the person from whose womb one is born, but each person's experience and emotional context can be entirely different.

Brain Network

Communication is based on expressive language and receptive understanding. As previously mentioned, expressive language is the function of Broca's area. Wernicke's area is responsible for understanding. Both areas are critical to communication.

Mirror neurons are a factor in communication. Also mentioned earlier, mirror neurons produce a form of empathy in which the receiver replicates stimuli from the sender in their own brain and understands their experiences. Reciprocally, the sender must be sensitive to the receiver.

When coming in contact with new information, the brain searches for a previously structured neural network in which the new information will fit. "If the fit is good, what was learned/stored previously gives meaning to the new information."[13]

Movement and Dance Connections

Movement is a nonverbal, symbolic language communicating feelings and ideas that reflect context. Ballet and modern dance have narrative and aesthetic intent. Cultural and tribal dance forms embody symbolism and meaning. Recreational or entertainment dance encompasses cultural values and beliefs communicated through their movements.

The bridge in communication that produces action is sometimes referred to as "transformation."[14] In dance, learning new techniques involves a feedback loop. The dancer perceives an action and transforms it to his or her body through the function of mirror neurons.[15] Self-reflection or outside feedback by the teacher provides new sensory information to refine movements. Choreographers and dance teachers use good verbal skills, a strong movement vocabulary that relates to each learner, and a variety of learning domains. Mastery is built through verbal explanations accompanied by physical and spatial demonstrations.

Movement becomes a "dance vocabulary." Just as different cultures have different languages, cultures reflect different movement vocabularies. However, movement has a primal base understood by humans across all cultures, making it a universal form of communication.

Movement Explorations

Creating: (a) Decide on an artistic intent to communicate. (b) Create a movement phrase that expresses your idea.

Performing: (a) Practice your phrase to enhance expressive dynamics. (b) Perform your phrase for someone else.

Responding: (a) Reflect on your movement and whether there are elements of it that you could communicate more strongly. (b) Receive feedback from another person. (c) Discuss what the phrase communicates and how it achieves the artistic intent.

Connecting: (a) Does the phrase relate to meaning or contexts from personal life or other contexts? (b) Did the other person relate to the same contexts, or did he or she experience other meanings?

COLLABORATION

Collaboration is not merely a skill for learning or work; it is a human imperative. Collaboration extends the horizon of knowledge and actions beyond personal limitations to a wider collective endeavor. All cultures value relationships and friendships. In a world that is increasingly connected through technology, working collaboratively is growing as a way to achieve goals.[16]

Most species have some type of social network. Collaboration is not unique to humans. Wolves hunt in packs. Insects such as ants and bees have specific jobs within complex social networks. Perhaps it is fostered in animals based on survival needs, and particularly in mammals due to the close connection with a mother. Verbal language makes humans unique and heightens capabilities for collaboration.[17]

The traditional business model is based on a dictatorial boss who commandeers workers according to policies set by management. As Americans have become more educated, this model has been challenged by workers who have greater ability to make informed choices and insist on having a voice in decision making. A more democratic, collaborative model has replaced the demagogic, authoritarian one.[18]

Brain Network

The human brain is wired to "connect."[19] Use of fMRI imaging reveals how the brain responds to social encounters and provides understanding of mental mechanisms that drive social behaviors, knowledge critical to improving individual lives and organizations.[20]

From almost the moment of birth, a "default network" functions in the brain when we have finished an action or come to a place of rest. It involves the same network as social connections. "Default network activity during rest may reflect an evolved predisposition to think about the social world in our free time rather than its being merely a moment-by-moment personal choice."[21] The continual return of the brain to a "social cognitive mode" explains the human need for social contact.

Classroom Applications

In collaborative situations, students feel valued, connected, and responsible for participating in the learning of others. Research suggests students who work together develop self-understanding, commitment, better performance, and feelings of belonging. Students who have a sense of belonging to a school community have fewer discipline problems and lower levels of alcohol or marijuana use, early sexual behavior, and emotional distress or suicide.[22]

A strong classroom community improves academic achievement. A study of students in six elementary schools measured their sense of classroom community and found students who experienced greater communal connections exhibit higher academic motivation and performance, like school more, show greater empathy, are more likely to help others, and display an ability to resolve conflicts. A high sense of community also levels the playing field for at-risk children.[23]

Collaborative projects have become mainstream pedagogy through discussions as students test, broaden, reinforce, and integrate their learning. Arts-based teaching strategies promote collaborative learning experiences.[24]

Movement and Dance Connections

Dance is a social art form that is mainly practiced in groups or communities. It is believed that every culture has a form of dance. While music lessons tend to be solo, dance classes are a group phenomenon. Most dances are performed with an ensemble, and most societies dance communally for ritual, artistic expression, or recreation. Dancers share space, move in a rhythm together, and often need to unify their movements.

Collaborative choreography is becoming prevalent in contemporary dance groups. Cooperative dance companies, dance "collectives," and nuclei of choreographers are now common. Hierarchies have leveled, and more democratic choreographic processes are used. Many well-known choreographers develop dances through integration of cast improvisation and their artistic intent. Collaboration as a means of creative invention widens possibilities and invites innovation.

Movement Explorations

The explorations in this section require working with other colleagues. Collaboration is by necessity a group activity.

Creating: Each person needs to develop his or her own movement sequence of at least five or six different actions. You may use a movement sequence you previously developed in this chapter or create a new one.

Performing: Each person performs his or her sequences for the other participants.

Responding: (a) Analyze the sequence created by each person. (b) Combine everyone's movements into one long sequence by reflecting on and manipulating movement transitions, spatial relationships, rhythms, and dynamics. (c) Reflect on the relationships developed in the collaboration.

Connecting: Does the integrated sequence have any contextual connotations, symbolism, or connections to each participant or collective experiences?

CREATIVITY

Arts education is assumed to foster creative thinking and innovative ideas. This overlooks the fact that a great amount of arts education is taught by imitation; reading notes but not understanding music or composing it; or replicating movements, but not understanding their compositional structure or meaning. This type of pedagogy is purely technical and akin to crayoning in a coloring book. It does not develop innovative thinking necessary in the twenty-first century.

This book begins and concludes with a discussion of creativity, coming full circle. Creativity stands atop the revised Bloom's pyramid of cognitive domains because it requires a synthesis of all higher-level thought processes. When the arts or any disciplines are taught by focusing on creative process as applied skill, students reach the pinnacle of cognition.

Brain Network

Creative activity uses executive functions of the brain that occur in brain synthesis and orchestration. Neuronal connections determine how or when they fire in response to stimuli and needs.[25] The hippocampus is involved with memory, the thalamus serves in decision making and sometimes inhibits behaviors, and Broca's area is a center for expressive language and the ability to abstract, use symbols, and categorize. The list is long, but the interplay between all parts of the brain is constant.

Classroom Applications

An individual's thoughts define the "self" as a uniquely dynamic self-organizing system. We are wired as creative thinkers and problem solvers. However, humans learn to conform thoughts and actions through education from parents, schools, and society. In essence, we are taught not to be creative by inhibiting innate creativity.[26]

Neural connections are enriched when exercised. The creative mind forges connections in new and different ways. Sometimes it means following an impulse that leads in new directions, experiencing a flash of insight that seems to come from nowhere, or finding a solution to a problem that has incubated and stewed. Free-floating original thoughts occur when multiple brain regions interact in the highly developed association cortex.[27]

Movement and Dance Connections

Dance can be taught as replication and repetition, or as creative exploration of the body moving in space and time through experimentation with energies and relationships. Dance is considered a creative art form, but most training is limited to conditioning in order to shape the body as an instrument capable of every possible movement. Even modern dance classes usually focus on technique rather than movement exploration or creative problem solving. College and university programs teach choreography in "composition" classes, but studios rarely give students opportunities to create dances. Those who become choreographers often do so in spite of rigid training.

Improvisation is a skill. The stimulus can come from music, a feeling, idea, experience, or the environment. Dance can be inspired by anything. Discovering a personal, authentic movement vocabulary is akin to a process of self-awareness. When adapting codified movements such as those from world dances, each individual combines and expresses movements uniquely.

To develop a dance, a choreographer starts with an idea or stimulus—an environment, issue, emotion, event, experience, story, or piece of music that is transferred (transformed) into movement. Students who only learn specific steps and have never explored movement usually need guidance through a transference process before they can find a personal, creative dance vocabulary.

In a quantitative study, high school students who took dance in school were compared to a control group not enrolled in high school dance classes. Control students currently studying dance outside school were eliminated. The TTCT, the test administered to measure creativity, used drawings to measure fluency (number of ideas expressed), originality (unusualness of responses), abstractness of titles for drawings, elaboration (number of pertinent details in drawing), and resistance to premature closure (not closing off a figure in the most direct way). Students with greater experience in a school dance program had higher scores for originality.[28]

The Root-Bernsteins analyzed Graham's creative process using her comments and notes in relation to their thirteen thinking tools on highly creative people. They found all twenty-first-century skills embedded in Graham's choreographic process.[29]

Human progress is based on the creative spirit. The arts reflect the aesthetic legacy of the human mind.

Movement Explorations

You have been creating movement explorations throughout the entire book. Return to any explorations that particularly inspire you. This final series is the culmination or cornerstone of the book.

Creating: (a) Find a movement source or think of an idea, feeling, theme, or artistic intent to express and communicate. (b) Create, select, and organize movements and structure a sequence or short dance study based on your idea and intent. (c) Revise your movements as needed after analysis or reflection.

Performing: (a) Execute the movements created in space, rhythmically and dynamically. (b) Practice your movement sequence or short dance study for accuracy and expression.

Responding: (a) Analyze your movement sequence or short dance study based on use of the elements of space, time, energy, and movement patterns developed. (b) Interpret the meaning or context of your dance sequence or study. (c) Reflect and evaluate the choreography and performance, and revise your work to clarify and intensify your performance and communication.

Connecting: (a) Does your sequence or short dance study have personal meaning or relate to personal contexts in any way? (b) How does your sequence or short study relate to cultural or historical contexts? (c) Does your sequence or dance study relate to other aspects of knowledge?

NOTES

1. Rima Faber, "The Pedagogic and Philosophic Principles of the National Standards for Dance Education" (PhD diss., American University, 1997).

2. Commission on the Skills for the American Workforce, *America's Choice: High Skills or Low Wages!* (Washington, DC: National Center on Education and the Economy, 1990).

3. New Commission on the Skills of the American Workforce, *Tough Choices, Tough Times* (Washington, DC: National Center on Education and the Economy, 2007).

4. "P21 Partnership for 21st Century Learning," accessed June 6, 2015, http://www.p21.org/; "Framework for 21st Century Learning," accessed June 6, 2015, http://www.p21.org/our-work/p21-framework.

5. "Framework for 21st Century Learning"; Barbara Rich and Johanna Goldberg, *Neuroeducation: Learning, Arts, and the Brain* (New York/Washington DC: Dana Press, 2009).

6. Mariale M. Hardiman, "The Creative Artistic Brain," in *Mind, Brain, & Education: Neuroscience Implications for the Classroom*, ed. David A. Sousa (Bloomington, IN: Solution Tree Press, 2010), 227–46.

7. George McCloskey, "Executive Functions in the Classroom" (workshop, Learning and the Brain, University of Maryland, College Park, MD, November 17, 2014).

8. Jerome Kagan, "Why the Arts Matter: Six Good Reasons for Advocating the Importance of Arts in School," in *Neuroeducation: Learning, Arts, and the Brain*, eds. Barbara Rich and Johanna Goldberg (New York/Washington, DC: Dana Press, 2009).

9. Hardiman, "The Creative Artistic Brain."

10. Nancy Lampert, "Inquiry and Critical Thinking in an Elementary Art Program," *Art Education* (November 2013): 6–11.

11. Rima Faber, "Standards for Learning and Teaching Dance in the Arts and 21st Century Skills" (paper, Silver Spring, MD, 2005).

12. "National Core Arts Standards in Dance, Introduction for Dance," accessed January 3, 2015, www.nationalartsstandards.org.

13. Patricia Wolfe, *Brain Matters: Translating Research into Classroom Practice* (Alexandria, VA: Association for Supervision and Curriculum Development, 2001), 72.

14. James E. Zull, *From Brain to Mind: Using Neuroscience to Guide Change in Education* (Sterling, VA: Stylus Publishing, 2011).

15. Emily S. Cross, "Building a Dance in the Human Brain," in *The Neurocognition of Dance: Mind, Movement and Motor Skills*, eds. Bettina Blasing, Martin Puttke, and Thomas Schack (New York: Psychology Press, 2012); Scott Grafton and Emily Cross, "Dance and the Brain," in *Learning, Arts, and the Brain: The Dana Consortium Report on Arts and Cognition*, eds. Carolyn Asbury and Barbara Rich (New York/Washington, DC: Dana Press, 2008), 61–69.

16. Matthew D. Lieberman, *Social: Why Our Brains Are Wired to Connect* (New York: Crown Publishers, 2013).

17. Lieberman, *Social*.

18. Lieberman, *Social*.

19. Lieberman, *Social*, x.

20. Lieberman, *Social*.

21. Lieberman, *Social*, 20.

22. Pamela LePage, Linda Darling-Hammond, and Hanife Akar, "Classroom Management," in *Preparing Teachers for a Changing World: What Teachers Should Learn and Be Able to Do*, eds. Linda Darling-Hammond and John Bransford (San Francisco: Jossey-Bass, 2005): 327–57.

23. LePage, Darling-Hammond, and Akar, "Classroom Management."

24. Mariale Hardiman, "A View from Education," in *Neuroeducation: Learning, Arts, and the Brain*, eds. Barbara Rich and Johanna Goldberg (New York/Washington, DC: Dana Press 2009), 70–74.

25. Nancy C. Andreasen, *The Creative Brain: The Science of Genius* (New York: Penguin Group, 2005).

26. Andreasen, *The Creative Brain*.

27. Andreasen, *The Creative Brain*.

28. Sandra Minton, "Assessment of High School Students' Creative Thinking Skills: A Comparison of Dance and Non-Dance Students," *Research in Dance Education* 4, no. 1 (2003): 31–49.

29. Michele Root-Bernstein and Robert Root-Bernstein, "Martha Graham, Dance, and the Polymathic Imagination: A Case for Multiple Intelligences or Universal Thinking Tools?" *Journal of Dance Education* 3, no. 1 (2003): 18.

Glossary

acetylcholine—A chemically based neurotransmitter released at the connections between neurons, especially the connections with muscle cells.

action observation network (AON)—A network of brain cells activated when observing movement. Previous physical practice of the observed movements is necessary for robust activity in parts of the AON.

adrenal glands—An organic structure in the endocrine system sitting atop the kidneys in humans that releases hormones during times of stress.

adrenaline—One of the hormones secreted by the adrenal glands in reaction to stress that produces excitation. Also known as epinephrine.

alpha waves—Neural oscillations in the frequency range of 7.5 to 12.5 hz. Associated with a relaxed state and creative process.

AMP—A substance that has the capacity to amplify neuronal response during the formation of memories.

amygdala—A brain area concerned with emotions, especially fear; part of the limbic system.

anterior cingulate—Part of the brain's cortex that helps focus and awareness of thoughts.

aphasia—The inability to either express or understand language.

apraxia—The inability to perform learned movements despite having the physical ability, desire, and knowledge of expectations.

axon—An extension of a neuron that carries impulses away from its cell body.

basal ganglia—A group of cells found in the inner layers of the cerebrum of the brain that have a role in controlling movement and some cognitive functions.

brain stem—Three anatomical structures located at the bottom of the brain that process some sensations and mediate life support.

Broca's area—The brain area, located in the left frontal lobe, responsible for expressive language and the ability to use abstract thought.

Cash Multidimensional Body-Self Relations Questionnaire—A self-report inventory used to assess body image, especially attitudes toward one's physical self.

central nervous system (CNS)—The part of the nervous system found within the skull and vertebral column, consisting of brain and spinal cord.

cerebellum—Part of the brain located beneath the occipital lobe that is responsible for movement control, coordination, and learning new motor skills.

cerebral cortex—The outermost layer of the brain composed of the cerebral hemispheres responsible for higher brain functions such as executive functions, perception, nuanced emotions, abstract thinking, and planning.

chunking—Organizing information into clusters or patterns based on underlying strategies or functions.

consolidation—A process in which short-term memories are transferred to long-term memories or link meaningful associations with previous memories.

cortical homunculus—A pictorial diagram representing parts of the brain that process and integrate motor information and tactile sensations.

DNA (Deoxyribonucleic acid)—A long polymer made from repeating units called nucleotides that transmits genetic instructions in all living organisms.

dopamine—A neurotransmitter associated with satisfaction, pleasure, and paying attention. It is produced when engaging in physical activity and active learning.

empathizing—A reflective response to incoming stimuli producing vicarious experience or thoughts of another individual.

encoding—The process by which we interpret or analyze sensory input and store it.

endocrine system—A collection of glands that secrete hormones and enzymes into the blood.

endorphins (endogenous opioid neuropeptides)—A natural opiate produced by the central nervous system and pituitary gland that serve to reduce pain.

entrainment—The alignment of an organism to coordinate internal rhythms and actions to those in the environment.

epinephrine—A hormone secreted by the adrenal glands. Also known as adrenaline.

episodic memory—Memories connected to past, personal events.

explicit memory—The conscious recall of people, places, things, events, and facts. It is the same as declarative memory.

far transfer—Learning between contexts that seem unrelated.

fascia—A band or sheath of connective tissue that supports or connects muscles and/or organs in the body.

feedback loop—A neurological cycle that connects the motor and sensory systems in response to environmental changes.

flashbulb memory—The recall of emotionally charged events in explicit detail and reality.

foundation image—Common images in visual input such as a right angle that trigger neuronal firing patterns in the brain.

frontal lobe—A large part of the brain behind the forehead responsible for high-level cognitive processes such as planning, controlling emotions, and making decisions.

functional magnetic resonance imaging (fMRI)—An electronic brain-scanning technique that depicts brain activity in specific areas fueled by glucose and oxygen in the blood.

fusiform gyrus—The crest of a fold in the brain near the bottom, inner part of the temporal lobe that has parts for recognizing color, faces, and objects.

glucocorticoid—Substances that regulate a number of body processes such as metabolism and homeostatic functions.

glutamate—A neurotransmitter in the brain that plays a major role in exciting and linking neurons during learning and long-term memory.

golgi tendon organ—Proprioceptive organs located in muscles between their belly and tendon, providing information about pull on a tendon.

gyrus—The crest of a bulge or fold in the brain.

hippocampus—A brain structure located deep in the temporal lobe that is necessary for the formation of explicit memories.

hypothalamus—A part of the brain located below the thalamus that controls autonomic, endocrine, and visceral functions.

implicit memory—Memories that do not require conscious recall such as habits, perceptual, or motor strategies. Also known as procedural memory.

insula—A cortical area buried deep in the brain formed from parts of the frontal, parietal, and temporal lobes. It receives input from internal and sensory organs to determine gut feelings and disgust.

integrative cortex—The cortical process of gathering information with which we engage and attempt to understand.

intrafusal fiber—Muscle fibers located within muscle spindles that ensure a continuous flow of sensory information concerning muscular stretch, contraction, or possibly velocity to the central nervous system.

joint receptors—A part of the body's proprioceptive system located in joint capsules and ligaments that has a more limited contribution to kinesthesia than the muscle spindles.

kinase—A chemical in the brain that activates genes in preparation for forming long-term memories.

lateral pulvinar—A part of the thalamus that remains active while other parts of the brain are quiet waiting on alert in response to a stimulus that might be a threat.

limbic system—A group of brain structures that regulate emotion.

locomotor/nonlocomotor—Movements that travel across space, and those that do not travel.

long-term memory—A persistent strengthening of neurological patterns resulting in continued recall of events and experiences.

lumbar spine—Five vertebrae that have a forward curve in the lower part of the spine.

magnetoencephalography (MEG)—A brain-scanning technique that picks up the small magnetic pulse generated from neuronal activity.

memory extinction—A process and result occurring when recalled memories are contradictory and weaken as a result.

mental rotation test—Subjects view pictures of several two- or three-dimensional objects that have been rotated on an axis and must decide if the objects are the same even though their position has been changed.

meta-analysis—A statistical analysis in which the results of different studies are combined and compared to find similar patterns or disagreements.

metacognition—The act of being aware and expressing one's own thought process.

mirror neurons—Brain cells involved in empathy that become active when observing facial expressions, and energies from, or movements of another person.

motor cortex—An area of the cortex located in the back part of the frontal lobe that controls movement.

muscle spindle—Encapsulated structures scattered throughout all voluntary muscles whose fibers run parallel to the main muscle fibers, and which respond to stretching or lengthening of the muscle.

National Core Arts Standards in Dance—A grade-by-grade guideline from pre-K through grade twelve of what students should know and be able to do in dance that was released online to the public in 2014.

near transfer—Learning between contexts that seem related.

negative transfer—A process in which the brain seeks to fit new information into an existing brain network but cannot find a total fit—a situation that hinders learning.

neural impulse—When a nerve cell actively fires.

neural map—The pattern of connections between parts of the central nervous system.

neural network—A group of interconnected neurons.

neuro-education—Teaching strategies based on neuroscience knowledge of brain functions.

neurogenesis—The generation of new neurons.

neurological pattern—A habitual recurrence of activity between connected neurons related to specific functions.

neurotransmitter—A chemical released at the connection between neurons facilitating transmission.

noradrenaline—A chemical in the brain that incites mental arousal and an elevated mood. Also known as norepinephrine.

norepinephrine—A chemical in the brain that incites mental arousal and an elevated mood. Also known as noradrenaline.

nucleus accumbens—A part of the basal ganglia that has a role in motivation, pleasure, and learning.

occipital lobe—One of four main divisions of the brain located at the back for processing visual stimuli.

optic chiasma—A structure by which electrical pulses generated by visual stimuli cross to the opposite side and reverse the visual image.

parietal lobe—One of the parts of the cerebral cortex located at the top and back of the brain that processes spatial input and performs sensory integration.

perceptual motor skills—Physical abilities that depend on sensation, perception, and planning.

phonological loop—Part of short-term or working memory that stores sound information.

pituitary gland—A small gland about the size of a pea located at the base of the hypothalamus almost at the center of the brain. It secretes hormones that control a variety of body processes, including metabolism.

place cells—A map of external space created in the pyramidal cells of the hippocampus.

plasticity—The ability to adapt when there is a loss of function or change of environment.

positive transfer—A process in which the brain seeks to fit new information into an existing brain network and finds a good fit—a situation that assists learning.

positron emission tomography (PET)—A brain-scanning system that locates the most active brain areas based on intake of fuel. It requires the injection of a radioactive marker, a major drawback.

precentral gyrus—One of four gyri of the frontal lobe that is part of the motor cortex.

prefrontal cortex—The forward part of the frontal cortex involved in planning behavior and memory.

premotor cortex—A part of the frontal cortex that plans and creates movements.

procedural memory—The type of memories that do not require conscious recall such as habits, tasks, and perceptual motor strategies. Also known as implicit memory.

proprioception—Sensory information (joint position, muscular forces, orientation) sent from within the body.

pruning—A process in a brain in which neurons search for other neurons to which they can link. If a neuron does not find its place in the scheme of things, it dies or is pruned.

pyramidal cells—Hippocampal cells that create a map of external space. Also known as place cells.

Radell Evaluation Scale—An assessment instrument to measure growth in movement ability.

reconsolidation—A process in which recalled memories are checked and become stronger if they prove to be correct.

reticular activating system (RAS)—A group of nuclei at the top of the brainstem that trigger arousal, the first stage in paying attention.

retrieval—The act of recalling information from memory.

reward center—Human pleasure reactions connected to a number of distributed brain regions, especially the limbic system.

right hemisphere—The right half of the brain.

schema—An underlying structure connecting information together in the mind based on a relationship to a central theme.

semantic memory—A form of memory based on the meaning of an object, event, or concept.

semicircular canal—Three semicircular structures oriented at approximate right angles to each other located inside each ear. They respond to changes in head movements, inform the body's spatial orientation, and stabilize balance.

serotonin—A neurotransmitter that regulates moods, including depression, anxiety, food consumption, and violence.

short-term memory—Information retained in the mind for only seconds to minutes.

somatosensory cortex—The portion of the cerebral cortex, located in the parietal lobe, that processes sensations, including touch, vibration, pressure, and sense of limb position.

striatum—Part of the basal ganglia important to movement and cognition.

superior colliculus—Part of the mental apparatus located deep in the brain. It searches for and tracks movements and is attuned to sounds.

supplementary motor cortex—A part of the cortex located in front of the primary motor cortex that contributes to movement control, such as posture and movements internally generated.

synapse—A gap between adjacent neurons through which impulses are transmitted.

synovial fluid—A substance found inside synovial joint capsules that reduces friction.

temporal gyrus—The crest of a bulge or fold on the temporal lobe.
temporal lobe—One of the brain lobes located in the lower part of the brain near the ears. It is responsible for auditory processing and some types of memory.
terminal—The ending of an axon.
thalamus—A processing point for sensations and location for relaying motor information to muscles.
theory of mind—An ability to see the world from the viewpoint of another person and create a mental model of their thoughts and intentions.
transduction—The use of a virus to transfer DNA between cells.
transference/transformation—The application of a concept from one type of content or discipline to another.
Uses of Object Task—An assessment of the number of ways a designated object can be used.
vestibular apparatus—A sensory system located in the inner ear responsible for balance and spatial orientation.
visuo-spatial sketch pad—A part of working memory that caters to visual and spatial input.
Wernicke's area—Portion of the left parietal lobe concerned with language comprehension.
working memory—A type of short-term memory that integrates moment-to-moment perceptions over a relatively short period and combines them with memories of past experiences.

Index

21st Century Skills, 151, 153, 154, 159;
    4 C's skills, 152–59

Abstraction, 13;
    Transform, 14, 16
aging, 77
Ailey, Alvin, 106;
    Revelations, 106
Alexander, Frederick Matthias, 6
Alzheimer's disease, 77
anatomical terms, 163;
    adrenal glands, 163;
    fascia, 164;
    ligaments, 65
ancient Greece, 139;
    Socrates, 55, 139
arts-based teaching, 157
Arts Education Partnership, 136
assessment instruments & techniques:
    Cash Multi-Dimensional Body-Self
        Relations Questionnaire, 164;
    Effort Shape, 53;
    Laban Movement Analysis (LMA),
        6, 7, 8;
    meta-analysis, 128, 166;
    Metropolitan Achievement Test, 119;
    qualitative research, 144;
    Quick Neurological Screening
        Test II, 46;
    Radell Evaluation Scale, 168;
    Torrance Test of Creative Thinking
        (TTCT), 140;
    Use of Objects Task, 169
Association, 127;
    analogy, 14, 127, 128, 130;
    metaphor, 102, 127, 128, 129, 133;
    simile, 127, 128, 129
awareness: 3, 6, 10, 12, 14, 67, 85–86,
    108, 146–47
    spatial, 3, 12

Bartenieff, Irmgard, 6, 7, 9
Bloom, Benjamin, 51, 54, 56, 60n1,
    61n2, 158
brain anatomy, 1–4;
    anterior cingulate, 39, 94, 114, 163;
    amygdala, 63, 64, 65, 66, 67, 68, 78,
        83, 91, 114, 120, 153, 163;
    basal ganglia, 13, 45, 78, 94, 153,
        163, 167, 168;
    brain stem, 3, 39, 40, 66, 144, 163;
    Broca area, 4, 5, 38, 51, 54, 55, 94,
        120, 132, 154, 158, 164;
    Cerebellum, 3, 13, 16, 45,
        78, 83, 94, 110,
        122, 164;
    Hippocampus, 3, 65, 68, 78, 83, 88,
        91, 120, 158, 165;

Homunculus, 111, 164;
Insula, 65, 66, 68, 165;
integrative cortex, 124, 126, 165;
lateral pulvinar, 41, 166;
left hemisphere, 5, 51, 126, 132;
limbic system, 3, 63, 64, 78, 152, 166;
mirror neurons, 14, 23, 31, 32, 71, 72, 122, 154, 166;
motor cortex, 110, 166;
occipital lobe, 3, 22, 52, 83, 105, 132, 167;
place cells, 88, 167;
prefrontal cortex, 28, 39, 45, 52, 65, 71, 94, 167;
premotor area, 24, 31, 122, 153, 167;
pyramidal cells, 83, 167, 168;
recticular activating system, 40;
reward center, 45, 168;
right hemisphere, 10, 126, 168;
somatosensory cortex, 65, 110, 168;
striatum, 83, 168;
superior colliculus, 22, 39, 41, 168;
temporal gyrus, 25, 54, 168;
temporal lobes, 3, 165;
thalamus, 3, 13, 41, 94, 158, 169;
upper gyrus, 59;
Wernicke area, 54, 120, 154, 169
brain energy sources: 37–38;
glucocorticoid, 84, 165;
glucose, 37;
oxygenation, xviii, 37
brain functions, xvii, xviii, 4, 9, 13, 16, 38, 69, 77, 81;
association, 78, 83, 102, 125, 126, 127;
attention, 3, 16, 39–47, 65, 67, 81, 122, 140;
attention span, 44;
consciousness, 3, 39, 78, 85–87;
emotion, 31, 63–72;
empathy, 14, 23, 71–74, 154;
engagement, 37, 43, 44, 45, 46;
executive function, 39, 45, 47n1, 48n12, 59, 142, 152, 158;
imagination, 101–2, 122, 139;
memory, 3, 4, 40, 52, 77–88, 91–95;
plasticity, 57, 167;
problem solving, 13, 21, 41, 139–47;
theory of mind, 71, 72, 169;
thinking, 1, 3, 7, 13–16, 51, 54–56, 105, 119, 139–40, 148, 152–54;
unconsciousness, 86
brain/mental conditions:
amnesia, 91, 94;
anorexia, 114, 115, 118n49;
aphasia, 4, 163;
Asperger's Syndrome, 72;
Austism, 66, 72;
bulimia, 114, 118n49;
repression, 94
brain waves: 102
alpha waves, 163;
entrainment, 164
Broca, Paul, 2, 4, 5, 18n8
Bush, George H. W., 151
Bush, George W., 151

cardiorespiratory endurance, 38
cells, building blocks of, 1;
adenine, 1;
cytosine, 1;
guanine, 1;
thymine, 1
choreographic terms, 29;
ABA, 29;
Canon, 29;
Composition, 56;
dance making, 43, 67, 102, 144, 154;
dynamics, 8, 29, 104, 143, 157;
energy, 8, 43, 54, 72;
improvisation, 14, 45, 60, 102, 104, 143, 159;
improvise, 15–16, 82, 141, 153;
inspiration, 13, 26, 46, 60, 102, 141;
intent, 11, 60, 82;
narrative dance, 155;
pathway, 26, 29, 54, 89, 125, 130;
pattern, 9, 14–15, 25, 29, 104, 123–25;
phrase, 82, 85, 125–27, 129;
repertory, 23, 143;

rhythm, 8, 26, 29–30, 53, 124–25, 157;
rondo, 29;
sequence, 12, 14, 29, 43–44, 53, 57, 80–82, 90, 95–96, 157, 160;
space, 8–10, 53, 85, 125;
study, 143;
time, 8
classifying thinking skills:
Anderson and Krathwohl's Taxonomy, 51, 61n2, 61n20;
Bloom's Taxonomy, 51, 56, 61n20
cognition: 10, 13, 39, 51, 54, 78, 121, 140, 152–53
categorization, 51, 52;
consciousness, 78, 82, 85, 86, 87;
critical thinking, 21, 54, 55, 56, 151, 152, 154;
elementary school thinking;
embodied cognition, 39, 121;
forming patterns, 13, 14, 15;
high level thinking skills, xviii, 38, 51, 54, 133;
high school thinking, 154;
logic, 11, 55, 101, 139, 142–43;
mental bias, 24;
metacognition, 11, 60, 62n36, 85;
middle school thinking, 52, 154;
mind maps, 14, 85, 88, 90, 147;
neural patterns, 64, 102, 124, 130;
organization, 51–53, 110;
pattern formation, 78;
pre-K thinking, 154;
reflection, 41, 60, 146, 147;
schema, 93, 153, 168;
schemata, 153;
thinking tools, 13, 14, 16, 33n11, 159;
transference, 130, 131, 132
Cohen, Bonnie Bainbridge, 6, 18n8
Commissions & publications:
Commission on the Skills of the American Workforce, 151, 160n2;
New Commission on the Skills of the American Workforce, 151, 160n3;

Tough Times, Tough Choices, 160n3
Confucius, 120
Creativity:
Aha experience!, 141, 149n23;
Curiosity, 40, 41, 43, 120, 121;
divergent thinking, 139, 140;
flash of insight, 158;
flow, 124;
free association, 101;
incubation, 101;
inspiration, 13;
stimulus, 2, 37, 64, 103, 159
Cunningham, Merce, 33;
Points in Space, 33

dance genres:
aerobic, 37, 38;
ballet, 8, 53, 67, 79, 84, 106, 115, 127, 131, 133, 155;
flash mob dance, 43;
hip hop, 53, 120, 121, 124, 125;
jazz dance, 125, 127;
modern dance, 34n19, 53, 57, 79, 84, 106, 155;
pantomime, 131, 141;
tap dance, 29;
tribal dance, 53, 106, 124, 155
dance integration, 16, 122;
geometry, 120, 130;
history, 132;
language arts, 16, 119;
math, 132, 142;
science, 120;
Science with Dance in Mind, 120, 130, 135n9;
social studies, 131
dance movements:
axial movements, 53;
locomotor movements, 53, 166;
nonlocomotor movements, 53, 166;
dance teaching:
proprioceptive feedback, 26, 107
dance terms:
alignment, 104;
pattern, 25, 29–30, 123–25;

rhythm, 8, 53, 124, 143;
technique, 43, 46, 56, 58, 72, 104, 122, 141, 143, 147;
Delsarte, Françoise, 92
Dewey, John, 41, 86, 119, 122, 134n2,
different types of knowledge:
   conceptual, 42;
   factual, 54
diversity, 114
DNA, 1, 121, 164
Duncan, Isadora, 5, 14, 67, 92
Dunham, Katherine, 67;
   Southland, 67

effects of stress on learning, 68–69
embody, 121, 127
emotion, 63–69, 71, 91–92, 96, 140, 159
empathy, 14, 23, 71–72, 154:
   function of mirror neurons in, 74
endocrine system: 3, 164
   cortisol, 65, 68, 69;
   endorphins, 65;
   hormones, 64, 65, 164;
   pituitary gland, 3, 164, 167
engagement, 37, 43, 44, 45, 46, 134n5

Feldenkrais, Moshe, 5, 6, 7, 18n16
Freud, Sigmund, 86, 98n44

Gardner, Howard, 11, 18n25, 105, 116n7, 139;
   Multiple Intelligences, 11, 116n23, 148n1;
   Project Zero, 134n7
genes, 41, 79, 165
Graham, Martha, 13, 32, 35n39, 46, 57, 58, 67, 87, 104, 127, 159;
   Appalachian Spring, 13, 127;
   Lamentation, 32, 35n39
Grandin, Temple, 73n20, 105, 116n25
Green Gilbert, Anne, 9, 18n23;
   body-side movements, 9, 10;
   Brain Dance, 46;
   core-distal movements, 9;
   cross-lateral movements, 9, 10;
   head-tail movements, 9, 10;
   upper-lower movements, 9, 10;
   vestibular spinning, 10

Holm, Hanya, 102
Homeostasis, 65;
   state of balance, 68
Humphrey, Doris, 14;
Hutchinson Guest, Ann, 143, 148n18

imagination:
   auditory imagery, 103–5;
   body image, 113, 115;
   body schema, 113;
   direct imagery, 64;
   female body image, 113;
   indirect imagery, 64;
   kinesthetic imagery, 112;
   male body image, 113;
   mental imagery, 106;
   mind map, 22, 90, 105–6;
   tactile imagery, 112;
   visual imagery, 103, 105, 106, 112;
   visualize, 104, 105
Impressionism, 92
integrated arts, 84, 130
intelligence:
   interpersonal intelligence, 12;
   intrapersonal intelligence, 12;
   linguistic intelligence, 11;
   mathematical-logical intelligence, 11;
   musical intelligence, 11;
   naturalist intelligence, 13;
   spatial intelligence, 12, 105;
   Theory of Multiple Intelligences, 11, 116n23, 148n1
International Baccalaureate Dance Program, 73n24
intervention techniques, 46, 92;
   dance therapy, 92;
   Gestalt therapy, 92;
   Neurologcal Reorganization Therapy Program, 46, 49n48

Index

James, William, 39, 48n9

Kandel, Eric, 22, 78, 79, 33n5, 96n1, 97n12
kinesthetic sense:
    golgi tendon organ, 108, 109, 110, 117n34, 165;
    joint receptors, 108, 110, 165;
    muscle spindle, 108, 110, 165, 166;
    proprioception, 168;
    semicircular canals, 88, 168;
    skin receptor, 108, 110;
    vestibular apparatus, 88, 89, 110, 169

Laban, Rudolf, 6, 7, 8, 9, 18n22, 53, 143;
    Choreutics, 143;
    Effort Shape, 53;
    Eukinetics, 143;
    Laban Movement Analysis (LMA), 6, 7, 8;
    Labanotation, 8, 143, 148n18
Lerman, Liz, 121, 143;
    Ferocious Beauty Genome, 121
Lighthill, Sir James, 140
Limón, José, 14, 33;
    Moor's Pavane, 33
Luria, Alexander, 10, 132

McGregor, Wayne, 106
Memory:
    consolidation, 83, 95, 146, 164;
    cues, 92, 94, 95, 96;
    declarative memory, 83, 85;
    emotional memory, 83, 91, 92;
    encoding memories, 83, 98n60, 164;
    episodic memory, 86, 87, 164;
    explicit memory, 78, 83, 85, 164;
    extinct, 95, 166;
    flashbulb memory, 91, 92, 165;
    immediate memory (part of short-term), 81;
    implicit memory, 83, 85, 165;
    long-term memory, 81, 82, 83, 84, 85, 105, 142, 164, 166;
    motor memory, 84, 165, 167;
    phonological loop, 81, 167;
    plasticity of memory, 57;
    procedural memory, 79, 83, 165, 167;
    recall, 3, 64, 77, 83, 91, 92, 94, 95, 96;
    recognition, 94;
    reconsolidation, 94, 146, 168;
    reinforce memories, 95;
    retrieval, 77, 78, 80, 94, 95, 168;
    role of chunking, 93, 94, 95;
    role of practice & repetition, 29, 80;
    semantic memory, 86, 87, 168;
    short-term memory, 80, 81, 82, 85, 124, 142, 164, 168, 169;
    spatial maps, 112;
    spatial memory, 88;
    storage, 77, 78, 79, 80, 91;
    visuo-spatial sketchpad, 81, 169;
    working memory, 40, 52, 54, 69, 80, 81, 82, 140, 167, 169
mental imagery, 5, 23, 44, 102, 103, 116n12, 152
mind/body connection, 14, 69;
    felt-thought, 4;
    somatic systems, 5, 6, 7
motor conditions, 59;
    apraxia, 59, 163
motor system:
    fine motor skills, 5, 11;
    gross motor skills, 5;
    motor control, 78;
    perceptual motor skills, 83, 167
movement analysis, 6, 7, 8, 53, 143
music terms, 7, 8;
    pitch, 7, 31;
    syncopated, 124;
    tempo, 8

National Core Arts Standards in Dance, 153, 154, 161n12, 166;
    Connecting, 154;
    Creating, 154;
    Performing, 154;
    Responding, 154

neurons:
    axons, 2, 163, 169;
    dendrites, 2, 124;
    neural network, 27, 41, 83, 130, 152, 155, 166;
    patterns, 6, 25;
    stimulation, 78, 84;
    synapses, 1–2, 28, 68, 124
neuroscience, xviii, 4
neuroscience, use of technology in:
    brain imaging, xvii, 72, 86, 98n46;
    electroencephalography (EEG), 102, 140;
    functional magnetic resonance imaging (fMRI), 165;
    magnetoencephalography (MEG), 126, 166;
    positron emission tomography (PET), 55, 125, 167
neurotransmitters:
    acetylcholine, 140, 163;
    adrenaline, 91, 163;
    AMP, 80, 163;
    dopamine, 3, 27, 28, 41, 45, 68, 140, 164;
    epinephrine, 164;
    glutamate, 80, 84, 165;
    kinase, 80, 165;
    noradrenalin, 69, 91, 167;
    norepinephrine, 68, 167;
    serotonin, 41, 78, 80, 168
Nikolais, Alwin, 23, 102
nonverbal communication, xviii, 30, 31, 32, 33
Noyes, Florence, 5, 104

observation: 13, 21, 23–24, 27
    action observation network (AON), 24, 31, 163;
    recognizing patterns, 13, 15, 27
optic system, 21;
    occipital lobes, 3, 22;
    retina, 21, 22;
    visual field, 22, 23

perception, 13, 24, 27, 68, 78, 88, 103
perceptual conditions:
    Attention Hyper Activity Disorder (ADHD), 52
Perkins, David, 120, 134n7
Piaget, Jean, 10, 11, 18n24
Picasso, Pablo, 105
Pilobolus, 144
Pink, Daniel, 44, 49n36, 148n16
Primary Movers, xixn3, 120
problem solving:
    adaptability, 57, 59;
    creative problem solving, 21, 102, 139–44;
    curiosity, 120–21;
    deductive reasoning, 142;
    divergent thinking, 139, 140;
    flash of insight, 158;
    flexibility, 57, 58, 139, 142, 153;
    framework, 42;
    free association, 101;
    inductive reasoning, 142;
    inference, xiv, 142;
    logic, 139, 142, 143;
    motivation, 44, 45, 46;
    self-determination, 59;
    self-direction, 59, 60, 153;
    trial and error, 83, 120

Realism, 102, 131
Root-Bernstein, Robert & Michele, 13, 14, 18n28;
    creative thinking tools, 13, 14

Sachs, Oliver, 116n21, 117n45, 124, 136n30
sense of self, 88, 89, 90;
    allocentric, 89, 90;
    egocentric, 10, 88, 89, 90;
    orientation, 12, 29, 88, 90, 105;
    self-concept, 114, 49n29;
    self-esteem, 114
sensory modalities: 31, 84–85
    auditory sense, 25;

kinesthetic sense, 25, 26;
polysensual, 13;
smell, 103;
tactile sense, 9, 25, 110, 112, 164;
visual sense, 25, 103
sleep (importance of), 82;
somatics practices, 5, 6;
   Alexander Technique, 6, 18n14;
   Awareness through Movement, 6, 18n16;
   Bartenieff Fundamentals, 6, 7;
   Body-Mind Centering, 6, 7, 18n17;
   Feldenkrais Method, 7;
   Functional Integration, 6;
   Ideokinesis, 6;
   Repatterning, 6
spinal cord, 1, 3, 13
Sweigard, Lulu, 6, 18n13
Symbols, 132, 133, 134, 154;
   alphabets, 133;
   archetype, 133;

teaching aids:
   flow chart, 147;
   graph, 147;
   table, 147;
   time line, 147;
   Venn diagram, 147
teaching strategies & systems:
   abstract transfer, 132;
   active learning, 40, 56, 79, 84, 95, 119, 120, 121, 139;
   Basic Reading through Dance Program, 122, 135n20;
   best practices, 70;
   brain friendly teaching, 136n29;
   chunking, 93, 94, 95, 96, 124, 146, 164;
   collaborative learning, 70, 119, 156–57;
   embodied writing, 123, 135n22;
   experiential learning, 41, 43, 81;
   factory model of learning, 42, 119;
   far transfer, 164;
   inquiry-based teaching, 153;
   learning styles, 95;
   literal transfer, 131;
   mnemonics, 104;
   Montessori System, 25, 59, 60, 62n35;
   near transfer, 130, 166;
   negative transfer, 166;
   neuro-education, 152, 166;
   play, 13–15, 42, 58, 91, 94, 102;
   positive transfer, 130, 167;
   rote learning, 41, 56, 86;
   Teaching for Understanding, 120;
   Transfer, 130–32, 140, 159, 169;
   for use with elementary school students, 131;
   for use with high school students, 159;
   for use with middle school students, 154;
   for use with pre-K students, 154;
   Viconic Language Method, 105
Tesla, Nikola, 101
Tharp, Twyla, 58, 62n30
Todd, Mabel, 6, 17n1

Vygotsky, Lev, 10

Werner, Heinz, 10

www.ingramcontent.com/pod-product-compliance
Lightning Source LLC
Chambersburg PA
CBHW031552300426
44111CB00006BA/279